Infectious Diseases

LECTURE NOTES ON

Infectious Diseases

BIBHAT K. MANDAL
FRCP
Consultant Physician and Director

EDMUND G.L. WILKINS
FRCP MRCPath
Consultant Physician

EDWARD M. DUNBAR
FRCP
Consultant Physician

all of the Regional Department of Infectious Diseases and Tropical Medicine
North Manchester General Hospital
Manchester

RICHARD T. MAYON-WHITE
FRCP FFPHM
Consultant Public Health Physician
Department of Public Health
The John Radcliffe Hospital
Oxford

Fifth edition

b
Blackwell Science

Editorial Offices:
Osney Mead, Oxford OX2 0EL
25 John Street, London WC1N 2BL
23 Ainslie Place, Edinburgh EH3 6AJ
350 Main Street, Malden
MA 02148 5018, USA
54 University Street, Carlton
Victoria 3053, Australia
10, rue Casimir Delavigne
75006 Paris, France

Other Editorial Offices:

Blackwell Wissenschafts-Verlag GmbH
Kurfürstendamm 57
10707 Berlin, Germany

Blackwell Science KK
MG Kodenmacho Building
7–10 Kodenmacho Nihombashi
Chuo-ku, Tokyo 104, Japan

First published 1969
Second edition 1974
Revised reprint 1975
Third edition 1980
Fourth edition 1984
Fifth edition 1996
Reprinted 1999

Set by Excel Typesetters, Hong Kong
Printed and bound in Great Britain by
MPG Books Ltd, Bodmin, Cornwall

For further information on Blackwell Science, visit our website:
www.blackwell-science.com

DISTRIBUTORS

Marston Book Services Ltd
PO Box 269
Abingdon
Oxon OX14 4YN
(Orders: Tel: 01235 465500
Fax: 01235 465555)

USA
Blackwell Science, Inc.
Commerce Place
350 Main Street
Malden, MA 02148 5018
(Orders: Tel: 800 759 6102
781 388 8250
Fax: 781 388 8255)

Canada
Login Brothers Book Company
324 Saulteaux Crescent
Winnipeg, Manitoba R3J 3T2
(Orders: Tel: 204 224-4068)

Australia
Blackwell Science Pty Ltd
54 University Street
Carlton, Victoria 3053
(Orders: Tel: 03 9347 0300
Fax: 03 9347 5001)

A catalogue record for this title is available from the British Library

ISBN 0–632–03351–7 (BSL)
ISBN 0–86542–676–7 (IE)

Library of Congress
Cataloging-in-Publication Data

Lecture Notes on Infectious Diseases
Bibhat K. Mandal . . . [*et al.*]
–5th ed.
p. cm.
Rev. ed. of: *Lecture Notes on the Infectious Diseases*/
Bibhat K. Mandal, Richard T. Mayon-White
4th ed. c.1984
Includes bibliographical references and index.
ISBN 0-632-03351-7
1. Communicable diseases.
I. Mandal, Bibhat K.
II. Mandal, Bibhat K.
Lecture notes on the infectious diseases.
[DNLM: 1. Communicable diseases.
WC 100 M271L L4713 1996]
RC111.L36 1996
616.9–dc20
DNLM/DLC 95–8411
for Library of Congress CIP

Contents

Section 4: Zoonoses, Tropical Diseases and Helminths

Preface to the Fifth Edition

This new edition differs radically from its predecessors both in the style of presentation and the content. Instead of concentrating on the infections known traditionally as 'infectious' or 'communicable' diseases, our aim has been to provide a practical guide to the diagnosis and management of clinical problems associated with infections on a much broader front. This has been achieved by combining disease-orientated descriptions with a clinical presentation/syndrome approach on a system-by-system basis. Microbial causes, discriminating features between similar diseases and basic principles of the management of important syndromes of infective aetiology have been dealt with in an easy-to-read format for each system. More detailed descriptions of the major infectious diseases follow, and include aspects of epidemiology and pathogenesis. Throughout the text, the more important pathogens have been highlighted in small capitals.

The rearrangement of presentation has allowed coverage of infections which were not included in the previous editions, e.g. infections related to the cardiovascular, skeletal and genitourinary systems, as well as to the eye, neonates and immunodeficiency states. The chapters on hospital infection and antimicrobial therapy have been completely rewritten. Important recent developments in the changing field of infection have received due attention. To bring about these changes E.G.L.W. and E.M.D. have joined B.K.M. and R.T.M.-W. as additional authors.

Despite the increase in the number of topics covered, our aim has been to keep the lecture notes to a modestly sized book. We hope it will continue to prove useful to medical students and young doctors in training, both in hospitals as well as in the community. Other health-care staff who are regularly involved with infection management, such as infection control and occupational health nurses, should also find the book of interest.

B.K. Mandal
E.G.L. Wilkins
E.M. Dunbar
R.T. Mayon-White

List of Abbreviations

AIDS	Acquired immunodeficiency syndrome
ALT	Alanine transaminase
AMP	Adenylate monophosphate
ARDS	Adult respiratory distress syndrome
ASD	Atrial septal defect
ASO	Antistreptolysin O
AST	Aspartate transaminase
BCG	Bacille Calmette–Guérin
CAH	Chronic active hepatitis
CBD	Common bile duct
CCDC	Consultant in Communicable Disease Control
CIE	Countercurrent immunoelectrophoresis
CK	Creatine kinase
CMI	Cell-mediated immunity
CMV	Cytomegalovirus
CNS	Central nervous system
CPH	Chronic persistent hepatitis
CSF	Cerebrospinal fluid
CT	Computerised tomography
CXR	Chest X-ray
DIC	Disseminated intravascular coagulopathy
DNA	Deoxyribonucleoside acid
dsDNA	Double-stranded DNA
EBV	Epstein–Barr virus
ECG	Electrocardiogram
ECM	Erythema chronicum migrans
EEG	Electroencephalogram
ELISA	Enzyme-linked immunoadsorbent assay
EM	Electron microscopy
ERCP	Endoscopic retrograde cholangiopancreatography
ESR	Erythrocyte sedimentation rate
FBCt	Full blood count
FTA	Fluorescent treponemal antibody
GBS	Guillain–Barré syndrome

HAV	Hepatitis A virus
HBV	Hepatitis B virus
HCV	Hepatitis C virus
HDV	Hepatitis delta virus
HEV	Hepatitis E virus
HHV-6	Human herpesvirus 6
Hib	*Haemophilus influenzae* type b (vaccine)
HIV	Human immunodeficiency virus
HLA	Human leucocyte antigen
HPV	Human papilloma virus
HSV	Herpes simplex virus
HTLV	Human T-cell leukaemia/lymphoma virus
ICP	Intracranial pressure
Ig	Immunoglobulin
IM	Intramuscular
IP	Incubation period
IV	Intravenous
IVDA	Intravenous drug abuse/abuser
LDH	Lactate dehydrogenase
LFTs	Liver function tests
LGV	Lymphogranulovenereum
LMN	Lower motor neuron
LP	Lumbar puncture
LVF	Left ventricular failure
MAC	*Mycobacterium avium* complex
MBC	Minimum bactericidal concentration
MDRTB	Multidrug-resistant *Mycobacterium tuberculosis*
MI	Myocardial infarction
MIC	Minimum inhibitory concentration
MMR	Measles, mumps and rubella
MR	Magnetic resonance
mRNA	Messenger RNA
MRSA	Methicillin-resistant *Staphylococcus aureus*
MSU	Midstream urine
MU	Megaunits
NA	Nucleic acid
NANB	Non-A non-B (hepatitis)
NNN	Novy–MacNeal–Nicolle (medium)
NSAID	Non-steroidal anti-inflammatory drug
OMPs	Outer membrane proteins
PAN	Polyarteritis nodosa
PCP	*Pneumocystis carinii* pneumonia

PCR	Polymerase chain reaction
PDA	Patent ductus arteriosus
PGL	Persistent generalised lymphadenopathy
PHN	Post-herpetic neuralgia
PID	Pelvic inflammatory disease
PML	Progressive multifocal leucoencephalopathy
PUO	Pyrexia of unknown origin
RNA	Ribonucleic acid
RSV	Respiratory syncytial virus
RTI	Respiratory tract infection
SLE	Systemic lupus erythematosus
SRSV	Small round structured virus
STD	Sexually transmitted disease
TNF	Tumour necrosis factor
TPHA	*Treponema pallidum* haemagglutination assay
TSS	Toxic shock syndrome
U&Es	Urea and electrolytes
UMN	Upper motor neuron
USS	Ultrasound scan
UTI	Urinary tract infection
VDRL	Venereal Diseases Research Laboratory
VHF	Viral haemorrhagic fever
VSD	Ventricular septal defect
VZV	Varicella zoster virus
WCCt	White cell count
ZN	Ziehl–Nielsen

Genus Abbreviations

Genus	Species
Aeromonas	*A. hydrophila*
Aspergillus	*A. funigatus*
Bacillus	*B. anthracis, B. cereus*
Bacteroides	*B. fragilis*
Bordetella	*B. parapertussis, B. pertussis*
Brucella	*B. abortus, B. melitensis, B. suis*
Campylobacter	*C. fetus, C. jejuni*
Candida	*C. albicans, C. glabrata, C. tropicalis*
Chlamydia	*C. pneumoniae, C. psittaci, C. trachomatis*
Clostridium	*C. botulinum, C. difficile, C. perfringens, C. tetani*
Corynebacterium	*C. diphtheriae, C. haemolyticum, C. ulcerans*
Coxiella	*C. burnetii*
Cryptococcus	*C. neoformans*
Entamoeba	*E. histolytica*
Enterococcus	*E. faecalis, E. faecium*
Escherichia	*E. coli*
Fusobacterium	*F. necrophorum*
Giardia	*G. lamblia*
Haemophilus	*H. ducreyi, H. influenzae*
Helicobacter	*H. pylori*
Klebsiella	*K. aerogenes, K. pneumoniae*
Legionella	*L. pneumophila*
Leishmania	*L. braziliensis, L. donovani, L. major, L. tropica*
Leptospira	*L. hardjo, L. icterohaemorrhagiae*
Listeria	*L. monocytogenes*
Moraxella	*M. catarrhalis*
Mycobacterium	*M. avium, M. bovis, M. leprae, M. tuberculosis*
Mycoplasma	*M. hominis, M. pneumoniae*
Neisseria	*N. gonorrhoeae, N. meningitidis*
Plasmodium	*P. falciparum, P. malariae, P. ovale, P. vivax*
Pneumocystis	*P. carinii*
Pseudomonas	*P. aeruginosa, P. pseudomanei*

Genus	*Species*
Rickettsia	*R. prowazekii, R. ricketsii, R. tsutsugamushi, R. typhi*
Salmonella	*S. enteritidis, S. paratyphi, S. typhi, S. typhimurium*
Schistosoma	*S. haematobium, S. japonicum, S. mansoni*
Shigella	*S. boydii, S. dysenteriae, S. flexneri, S. sonnei*
Staphylococcus	*S. aureus, S. epidermidis, S. saprophyticus*
Streptococcus	*S. milleri, S. pneumoniae, S. pyogenes, S. viridans*
Toxocara	*T. canis*
Toxoplasma	*T. gondii*
Treponema	*T. pallidum*
Trypanosoma	*T. brucei gambiense, T. brucei rhodesiense, T. cruzii*
Ureaplasma	*U. urealyticum*
Vibrio	*V. cholera, V. parahaemolyticus*
Yersinia	*Y. enterocolitica, Y. pestis*

SECTION 1

General Topics

CHAPTER 1

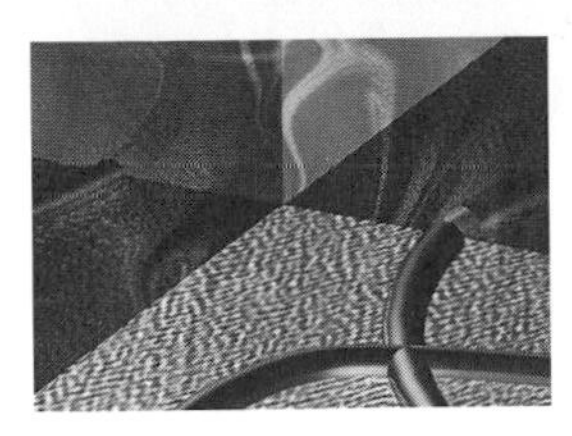

Introduction

CHANGING PATTERN OF INFECTIOUS DISEASES

During the last hundred years, there have been many environmental improvements in the developed countries of the world. These have included the provision of pure water supplies and the sanitary disposal of excreta, and improvements in food and milk hygiene, refuse disposal and the control of insects, lice and rodents. During the same period there have been great improvements in the nutrition of the population and in the development of health services, with the production of successful vaccines and effective antimicrobial drugs. The pattern of infectious diseases now encountered has changed dramatically as a result.

The current prevalence of infectious disease in countries like the UK can be summarised as follows.

- Infections which have been totally eradicated from the world: smallpox
- Infections which have virtually disappeared as endemic disease: cholera, typhus, diphtheria, poliomyelitis
- Infections which have become much less common or less virulent: measles, mumps, rubella, whooping cough, tetanus, tuberculosis, *Haemophilus influenzae* type b diseases, scarlet fever
- Infections whose incidence has remained unchanged: respiratory infections, chickenpox and zoster, hepatitis, infantile gastroenteritis (much less severe), infections of the nervous system (except *Haemophilus* meningitis), neonatal infections, urinary infections
- Infections which have increased: *Salmonella*, *Campylobacter* and other causes of food poisoning, sexually transmitted infections, infections in immunocompromised, debilitated and intensive care unit patients, *Clostridium difficile*, methicillin-resistant *Staphylococcus aureus* (MRSA) outbreaks of infection in hospitals
- Imported infections: malaria, enteric fever, amoebiasis, helminthiasis, exotic viral infections, traveller's diarrhoea

- New infection problems: acquired immunodeficiency syndrome (AIDS), multiresistance in pneumococci, salmonellae, tuberculosis, staphylococci.

TRANSMISSION OF INFECTION

Infection spreads by one of the following methods.

Airborne

Infection is exhaled from the case or carrier by coughing, sneezing or speaking in invisible respiratory droplets of moisture which are inhaled by the new host. The microorganisms may adhere to dust or textiles, leaving infected dust which may still transmit infection. Skin scales are an important source of contaminated dust. Dust may be carried by air currents, but rarely for distances of more than a few metres.

Diseases spread by airborne routes include:

Exanthemata: measles, rubella, chickenpox, scarlet fever

Mouth and throat infections: diphtheria, tonsillitis, mumps, herpes stomatitis

Respiratory tract infections: whooping cough, influenza and other respiratory virus infections, pulmonary tuberculosis

General: meningococcal and staphylococcal infection.

Intestinal

Infection present in the bowel excreta of a case or carrier is ingested by a fresh host. Transmission may be immediate and direct via infected fingers, eating utensils, clothing, toilets, etc., or indirect via food or water.

Diseases spread by the intestinal route include typhoid and paratyphoid, salmonellosis, dysentery, cholera, gastroenteritis, poliomyelitis and other enterovirus infections, viral hepatitis A and E.

In another group of ingestion diseases, transmission is direct from contaminated food. This group includes brucellosis, Q fever, salmonellosis, trichiniasis and other helminth infections.

Direct contact

Infection may be transmitted directly by local skin contact. These are mostly cutaneous infections and include impetigo and scabies.

Venereal route

Infection may be transmitted by sexual contact, including syphilis, gonorrhoea, lymphogranuloma venereum and herpes genitalis infection, human immunodeficency virus (HIV) and hepatitis B infection.

Insect or animal bite

Infections transmitted by bites include malaria, leishmaniasis, trypanosomiasis, typhus, rabies and simian herpesvirus infection.

Blood-borne

Some infections are commonly transmitted via infected blood or blood products, e.g. hepatitis B, HIV, hepatitis C.

These do not cover all the complex routes by which disease spreads. For example, leptospirae excreted in rats' urine may contaminate stagnant water and later penetrate the intact skin of a human host bathing in the water, or tetanus spores from the faeces of herbivorous animals may contaminate pasture land and years later may enter a wound and cause human disease.

Other diseases may spread by two or more alternative routes. For example, tuberculosis commonly spreads by airborne infection, but may spread via milk by ingestion or even by direct skin contact.

CONTROL MEASURES

Notification

In the United Kingdom, this system dates from 1899, and the list of notifiable diseases specified by the Public Health (Infectious Diseases) Regulations 1988 is detailed in Table 1.1.

NOTIFIABLE DISEASES	
Acute encephalitis	Paratyphoid fever
Acute policomyelitis	Plague
Anthrax	Rabies
Cholera	Relapsing fever
Diphtheria	Rubella
Dysentery (amoebic or bacillary)	Scarlet fever
	Smallpox
Food poisoning	Tetanus
Leprosy	Tuberculosis
Leptospirosis	Typhoid fever
Malaria	Typhus
Measles	Viral haemorrhagic fevers
Meningitis	
Meningococcal septicaemia (without meningitis)	Viral hepatitis
	Whooping cough
Mumps	Yellow fever
Ophthalmia neonatorum	

Table 1.1. Notifiable diseases

The law requires the doctor in attendance on the patient to notify the Consultant in Communicable Disease Control (CCDC) of the district in which the patient is living, as soon as the diagnosis is made. A local authority may make additional diseases notifiable in its own district.

Prompt notification enables the CCDC to instigate control measures where there is serious disease and also provides a valuable record of incidence. The Public Health Laboratory Service and other hospital laboratories are further important sources of information, from microbiological isolations. Other sources of information are nurseries, schools, large factories and selected 'spotter' general practitioners. In special circumstances, voluntary notification of a disease may be requested.

Communicable disease control

Outbreaks and serious cases of infectious disease are investigated by the CCDC, usually in collaboration with the Public Health Laboratory Service and infectious disease consultants. The source of infection, mode of spread, contacts and occupational circumstances are all investigated and appropriate measures carried out, including the isolation and treatment of patients and the immunisation and control of carriers and contacts. Food-borne outbreaks are usually investigated by environmental officers.

INTERNATIONAL CONTROL MEASURES

In cases of smallpox, cholera, plague and yellow fever (officially referred to as 'diseases subject to the regulations') the World Health Organisation arranges an interchange of information to enable the necessary public health and preventive measures to be carried out. The World Health Organisation also regularly exchanges information on a further group of infections which are kept under surveillance, and this includes poliomyelitis, epidemic influenza, louse-borne relapsing fever and louse-borne typhus fever.

CHAPTER 2

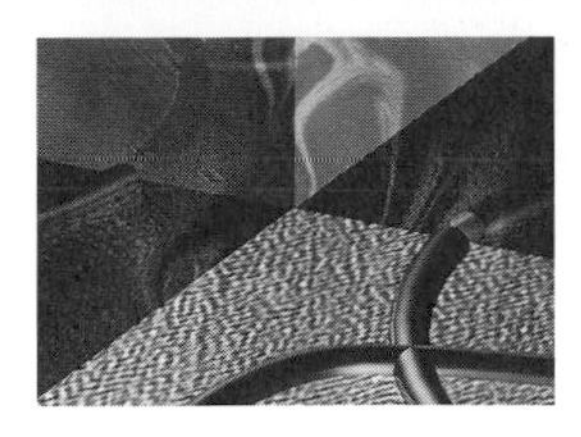

Immunology and Immunisation

IMMUNOLOGY OF INFECTION

The human body is continually exposed to a wide range of potentially pathogenic microbial organisms in its environment as well as within itself, yet most people do not experience recurrent or continued infections. This is due to the existence of a complex set of defence mechanisms.

Nonspecific defence mechanisms (active against any infective agent)

Prevention of entry

The *skin* and *epithelia* of the gastrointestinal, respiratory and genitourinary systems are effective barriers to microbial invasion. These barriers are supported by secreted *mucus* and *ciliary movement* of the respiratory tract, and by *gastric acidity* and *commensal flora*.

Nonspecific humoral mechanisms

These operate to eliminate microbes that evade the above barriers.

Complement system: this involves a group of plasma proteins which undergo a cascade of activation when triggered by an antigen (i.e. a microbe). The resultant production of active components help bacteriolysis and the antimicrobial activities of polymorphs and macrophages

Neutrophils: these migrate to the site of invasion and destroy microbes by phagocytosis

Mononuclear cells/macrophages: these are also able to ingest bacteria that have been coated with complement components or antibody

Natural killer lymphocytes: these can kill target cells without requiring antigenic or antibody stimulation and appear to help in resisting virus infections

Interferons: these are a group of glycoproteins which help to prevent viral replication in host cells

Inflammatory response: mediated by some of the above factors and other cytokines, this helps to contain further spread of infection. *Acute phase proteins* are a range of serum proteins which are produced in increased quantities during inflammation (e.g. C_3 and C_4 complement products, C-reactive protein, haptoglobin, α_1-antitrypsin, etc.) and are often used as markers of inflammation.

Specific defence mechanisms (active against a particular microbe)

There is a succession of complex interactions involving monocytes/macrophages, T-lymphocytes and B-lymphocytes.

Humoral immune response

T-cells secrete helper factors which activate specific clones of B-

INFECTION IN THE IMMUNOCOMPROMISED HOST

Type of defect	Common infection problems
Neutrophil dysfunction (qualitative or quantitative)	Recurrent septicaemia or localised infection due to streptococci, staphylococci, *Candida* septicaemia, aspergillosis. Response to antibiotics poor
Lack of antibody (humoral immunodeficiency)	Recurrent upper and lower respiratory infection. Recurrent boils, abscesses and cellulitis. Staphylococci and pneumococci commonly involved
Complement deficiency	Recurrent bacterial infection (C_3). Recurrent disseminated gonococcal and meningococcal infection (C_5, C_6, C_7, C_8)
Defective cell-mediated immunity	Severe or protracted infections due to measles, varicella, cytomegalovirus, herpes simplex virus, *Salmonella*, *Listeria*, mycobacteria, certain fungi and parasites (*Candida*, *Histoplasma*, *Coccidiodes immitis*, *Cryptococcus*, *Pneumocystis carinii*, *Toxoplasma*, cryptosporidium, microsporidium, *Leishmania*, *Strongyloides stercoralis*)

Table 2.1. Infection in the immunocompromised host

lymphocytes, bearing receptors for that particular antigen, into proliferation and transformation to antibody-producing cells (*plasma cells*)

Antibodies are immunoglobulins which combat infection in a number of ways: activation of complement helping phagocytosis and neutralisation of microbes. Immunoglobulin type M (IgM) antibody is produced initially (disappearing after several months), followed by a more pronounced and persistent production of IgG antibodies. IgA antibodies are secreted on mucosal surfaces. These are important first-line defences against future infection (coating microbes and preventing adherence to mucosal cells)

At the same time, greater numbers of *memory cells* (antigen-specific B-cells) are produced, so further antigen exposure provokes a quicker and more vigorous antibody response.

Cell-mediated immunity

Activated T-cells also produce cytotoxic cells which destroy infected cells, and cytokines which prevent microbes from replicating within cells. CD4-bearing T-cells (T-helper cells) help antibody synthesis and produce lymphokines which activate cytotoxic T-killer cells and produce inflammation which are characteristics of delayed hypersensitivity reaction. CD8-bearing cells (T-suppressor cells) suppress antibody production.

Infection in immunocompromised host

Components of the body's immune response mechanism may be absent or subject to malfunction. The consequences of this are detailed in Table 2.1.

IMMUNISATION

Immunisation is the induction of artificial immunity by giving preformed antibodies (i.e. as immunoglobulin) (passive immunisation) or by giving an antigen as a vaccine (active immunisation). The term *vaccination*, derived from the name of the protection against smallpox, is interchangeable with *immunisation*.

• Immunoglobulin used for passive immunisation is either nonspecific, collected from pooled human blood donations (e.g. normal immunoglobulin for protection against hepatitis A) or specific, formed from high-titre sera of humans (e.g. zoster immunoglobulin to prevent chickenpox) or animals who have been recently vaccinated or have had the natural infection. The protection given by immunoglobulin is short-lived (because of its gradual elimination from the body), varying from 1 to

4 months, according to the dose of immunoglobulin, the level of antibodies in it and the disease in question

• Active immunisation uses antigens which can be killed organisms (e.g. heat-killed typhoid vaccine), live organisms with low (attenuated) virulence (e.g. measles vaccine), inactivated bacterial products (e.g. diphtheria toxoid), or selected antigens of the particular organism (e.g. pneumococcal capsular polysaccharide). The antigens can be made by fraction of disrupted pathogens, or by genetic engineering. Most bacterial vaccines produce humoral immunity but the bacille Calmette–Guérin (BCG) vaccine produces cell-mediated immunity.

ROUTINE IMMUNISATION OF CHILDREN

Most countries have national programmes of vaccination for children. These programmes differ because of variations in the child health services of different countries, because of differing priorities, and because there has been some nationalistic reluctance to conform. The key objective of childhood vaccination programmes is to vaccinate as many children as possible. The current British vaccination schedule for children is shown in Table 2.2.

In some other countries, injected inactivated polio vaccines are preferred to the live oral vaccine, hepatitis B vaccine is given in the first year of life, BCG is given to all infants, and measles, mumps and rubella (MMR) vaccine is boosted later in childhood.

BRITISH VACCINATION SCHEDULE

Age	Vaccine
2, 3 and 4 months	Diphtheria, tetanus and pertussis vaccine Conjugated *Haemophilus influenzae* vaccine Oral polio vaccine
12–18 months	Measles, mumps and rubella vaccine
4–5 years	Diphtheria and tetanus booster Oral polio booster
10–14 years	Rubella vaccine (girls only) BCG vaccine
15–18 years	Diphtheria booster Tetanus booster Oral polio booster

Table 2.2. British vaccination schedule for children

TIMING OF VACCINATIONS

- The complexity of vaccination schedules, whether for children as described above or for travellers and other adults, arises from a wish to achieve the best immune response. However, this complexity can defeat the main objective of getting everybody vaccinated if the schedules are inflexible or not understood
- Vaccines that are given at birth are for infections that become an immediate risk – congenital hepatitis B when the mother is infected, and BCG in places of high tuberculosis prevalence. For other vaccines, a delay until the infant is 2 or more months old gives the immune system time to mature
- The killed organism vaccines generally require multiple injections to produce sustained immunity. The diphtheria and tetanus toxoid vaccines, whole-cell pertussis and typhoid vaccines and the hepatitis B vaccines are examples of this. Boosters are needed either to give life-long immunity, or to reinforce protection when facing a high risk of infection – hepatitis B vaccine in some occupations, tetanus vaccine after an injury, typhoid vaccine before travel abroad or diphtheria vaccine in an outbreak. Aluminium and other compounds act as adjuvants for killed bacterial vaccines by increasing the local inflammatory response
- Live vaccines typically need only a single injection, because multiplication in the body mimicks the natural infection without symptoms. In the case of polio vaccine, multiple doses are given to ensure that there is immunity to all three types of poliovirus and to minimise the effects of competition from other intestinal viruses. Live vaccines should be given together, or spaced at 2–3 week intervals to avoid interference from the immune response to one or other injection. BCG should also be spaced 3 or more weeks from other vaccinations because of temporary immunomodulation
- Purified antigens like pneumococcal, meningococcal and *Haemophilus* capsular polysaccharides can confer immunity by single injection. The immune response is better, especially in children under the age of 2 years, if the polysaccharide is conjugated to a protein, for example tetanus toxoid.

VACCINATIONS FOR TRAVEL, WORK AND OTHER PURPOSES

Whereas the vaccination of children should be a routine and universal procedure, vaccinations for travel and occupational risks should be selected. Recommendations vary according to the perceptions of, and attitudes to, risk and on the balance of benefits and costs (in which inconvenience is highly rated). Table 2.3 lists the vaccines that are available, with some possible uses.

OTHER VACCINES AND THEIR USES

Vaccine	Nature	Indications
Anthrax	Cell-free culture filtrate containing antigen	Workers exposed to imported animal hides and bones
Cholera	Whole cell, killed	Poor efficacy of short duration; no longer recommended
Diphtheria	Low dose toxoid	Travellers to Russia, Ukraine and contacts during outbreak
Hepatitis A	Inactivated virus	Frequent travellers to countries where hepatitis is common; immunoglobulin is an alternative for short-stay occasional travellers
Hepatitis B	Recombinant surface antigen	Most health-care workers, people going to live in high endemic countries, injecting drug users and the sexually promiscuous
Influenza A and B	Inactive viral components	People who are likely to have complications from influenza because of old age, diseases of the heart, lung or kidneys, immunosuppression or diabetes
Japanese B encephalitis	Formalin-inactive mousebrain-derived virus	Travellers who stay for more than 1 month during the rainy season in rural areas of Asia where this disease occurs in epidemics (usually areas where rice growing and pig farming coexist)
Meningococcal A and C	Inactive capsular polysaccharide	Travellers to places with epidemic group A meningococcal meningitis and for people in local outbreaks of either group A or group C disease
Pneumococcal	Inactive capsular polysaccharide	People with an increased risk of severe pneumococcal disease because of asplenia, diabetes, immunosuppression or diseases of the heart, lung or kidneys
Polio	Live attenuated	Boosters for travellers to countries where poliomyelitis is endemic and for contacts during outbreak
Rabies	Inactivated cell-culture-derived virus	People who have been bitten by rabid or possibly rabid animals and people who have an increased risk of such exposure

Continued

Table 2.3. Vaccinations for travel and other purposes

OTHER VACCINES AND THEIR USES

Vaccine	Nature	Indications
		through work with animals or travellers to remote areas in endemic countries
Tetanus	Formalin-inactivated toxin (toxoid)	Following tetanus-prone injury (if unimmunised or reinforcing dose >10 years previously)
Tick-borne encephalitis	Inactivated virus	Camping or walking in forests of north Europe in late spring/summer
Typhoid	Three types: oral attenuated live, purified Vi antigen and killed whole cell	Travellers to places where typhoid is endemic
Varicella zoster	Live attenuated	Children with immunosuppression
Yellow fever	Live attenuated	Travellers to African and South American countries where yellow fever virus exists

CONTRAINDICATIONS TO IMMUNISATION

There are few absolute, and many false, contraindications to immunisation.

- The one important contraindication to vaccination is a severe adverse reaction to an earlier dose: anaphylaxis, encephalitis or severe local inflammation
- Rubella vaccine is contraindicated in pregnancy although vaccination has been found not to cause fetal damage. Yellow fever or polio vaccines should not be withheld when there is risk of significant exposure
- A period of acute infection is not the best time to give a vaccine but the minor symptoms of common illnesses of childhood should not delay immunisation. Antibiotics are not a contraindication
- Live vaccines should not be administered to patients who are immunocompromised. However, MMR appears to be safe for human immunodeficiency virus (HIV)-positive children, whereas inactivated polio should be used instead of oral live vaccine
- Asthma, eczema and stable neurological diseases are not contraindications. Egg allergy is a contraindication for influenza vaccine, and previous anaphylactic reaction to egg contraindicates MMR, influenza and yellow fever vaccines.

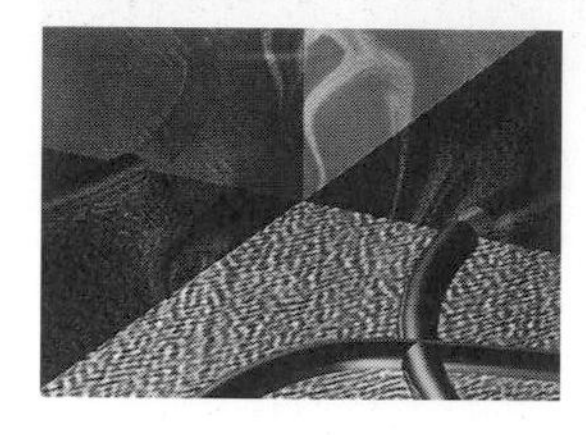

Chemotherapy

Antibiotics are the second most commonly used class of drug. One in three hospital patients receive antimicrobials, accounting for 25 per cent of total drug costs. One in twenty develop adverse reactions, occasionally severe. Their rational use, therefore, is important and requires knowledge of the infective causes of presenting syndromes, the spectrum of activity of antimicrobials, the principles of pharmacokinetics, the contraindications and interactions of drugs, and where to obtain assistance in making the choice (e.g. reference to the *British National Formulary*).

INAPPROPRIATE PRESCRIBING

Above all a doctor must be able to prescribe safely. Familiarity with a small list of antimicrobials is to be encouraged. In many situations antibiotics are not necessary, or are used inappropriately. Instances include:

- tonsillopharyngitis (usually viral)
- gastroenteritis (usually self-limiting if bacterial)
- wound infection (physical cleansing more important)
- where culture of an organism reflects contamination or colonisation but not infection (lack of knowledge of significant pathogens)
- multidose surgical prophylaxis (often continued beyond 24 hours).

Besides exposing the patient to the risk of unnecessary side effects, inappropriate antibiotic use will add to the selection pressure for resistant strains and may delay the start of the correct treatment.

Undertreatment is another problem. The most frequent errors of prescription are choosing the wrong drug and mistakes with antibiotic dose, duration or route of administration.

ANTIBIOTIC SELECTION AND USE

RANGE OF ANTIBIOTICS

A wide selection of antibiotics is available, and it is important to become familiar with the major representatives from each class.

Class (mode of action)	*Advantages/disadvantages*
β-lactams (inhibit cell-wall synthesis)	
Penicillins	
Benzylpenicillin (G), penicillin V	Safe, narrow spectrum
Ampicillin, amoxycillin	Narrow spectrum but including *Haemophilus influenzae*, enterococci
Amoxycillin–clavulanic acid (co-amoxiclav)	Broad spectrum includes *Staphylococcus aureus*, anaerobes and 'coliforms'
Flucloxacillin, cloxacillin (penicillinase-resistant)	Narrow spectrum but first-line drug for *S. aureus*
Azlocillin, piperacillin (ureidopenicillins)	Broad spectrum includes enterococci, anaerobes, 'coliforms' and *Pseudomonas aeruginosa*
Cephalosporins	
Cephalexin, cephradine, cefaclor (oral)	Modest Gram-positive, weak Gram-negative spectrum
Cefuroxime, cefotaxime, ceftriaxone (parenteral)	Broad spectrum includes *S. aureus*, streptococci, *Neisseria*, 'coliforms', *H. influenzae*
Ceftazidime (parenteral)	Excellent Gram-negative spectrum including *P. aeruginosa*
Monobactams	
Aztreonam	Excellent Gram-negative spectrum, hypersensitivity

Class (mode of action)	*Advantages/disadvantages*
	risk very low; inactive against Gram-positive organisms
Carbapenems	
Imipenem	Excellent Gram-negative, Gram-positive and anaerobic spectrum
Aminoglycosides (inhibit protein synthesis)	
Gentamicin, tobramycin, amikacin	Excellent Gram-negative spectrum including *P. aeruginosa*; synergistic with penicillin against enterococci and streptococci; no anaerobic activity
Quinolones (inhibit DNA replication)	
Ciprofloxacin, ofloxacin, norfloxacin	Excellent Gram-negative spectrum, modest Gram-positive activity; ciprofloxacin good for *P. aeruginosa*, enteric pathogens and atypical pneumonias
Macrolides (inhibit protein synthesis)	
Erythromycin, azithromycin, clarithromycin	Good Gram-positive spectrum; useful for penicillin-allergic patients and atypical pneumonia
Tetracyclines (inhibit protein synthesis)	
Tetracycline, oxytetracycline, doxycycline, minocycline	Modest broad spectrum but use limited to specific conditions (acne, *Chlamydia*, brucellosis, typhus, etc.); contraindicated in children and pregnant women
Sulphonamides and trimethoprim (inhibit folate synthesis)	
Sulphamethoxazole and trimethoprim (co-trimoxazole)	Modest broad spectrum; inactive against anaerobes; useful for chest and UTIs

Class (mode of action)	*Advantages/disadvantages*
Trimethoprim	Modest broad spectrum, useful only for UTIs
Glycopeptides (inhibit cell-wall synthesis)	
Vancomycin, teicoplanin	Excellent Gram-positive spectrum; inactive against Gram-negative organisms; vancomycin first-line drug for MRSA and *Staphylococcus epidermidis*
Miscellaneous	
Clindamycin (inhibits protein synthesis)	Good Gram-positive and anaerobic spectrum; risk of *Clostridium difficile*
Rifampicin (inhibits RNA synthesis)	First-line drug for tuberculosis; excellent Gram-positive, some Gram-negative activity (*Legionella*, *Neisseria meningitidis*)
Chloramphenicol (inhibits protein synthesis)	Good broad spectrum; risk of bone-marrow suppression
Fusidic acid (inhibits protein synthesis)	Narrow spectrum; additional drug for *S. aureus* infection
Metronidazole, tinidazole (inhibit DNA replication)	Excellent anaerobic and good protozoal spectrum

CHOOSING AN ANTIBIOTIC

In choosing an antibiotic, the doctor must:

- make a diagnosis as to the type of infection (e.g. pneumonia)
- have knowledge of both the organisms implicated and suitable antibiotics with activity against these organisms. The importance of whether an antibiotic is bactericidal or bacteriostatic is theoretical
- make a choice for the individual taking patient factors into account (e.g. pregnancy, age, renal or liver compromise, vomiting, allergy) and drug factors (e.g. penetration to site of disease, drug interactions)
- make use of the results of cultures and antibiotic susceptibility testing when they become available which will allow rationalisation of the em-

pirical therapy. Occasionally, the *in vivo* responses do not correspond, such as with aminoglycoside sensitivity with *Salmonella* (*in vitro* sensitive, *in vivo* resistant)

• take into account the need to monitor certain drugs where the toxic–therapeutic ratio is small (e.g. gentamicin, vancomycin).

The dose and route of administration must then be decided and an assessment as to the duration of total therapy made.

• Dose will depend upon weight, age, renal function, severity of infection, and toxicity of the agent (e.g. cefuroxime is preferred to gentamicin for severe Gram-negative infections in the elderly with impaired renal function)

• Route of administration depends mainly on the severity and site of infection; the presence of malabsorption, venous access and likely compliance may also be determining factors. If similar serum and tissue levels can be achieved by using high-dose oral antibiotic, there is no special need to continue intravenous (IV) antibiotics

• Duration of therapy rests on the response to therapy and likelihood of relapse (e.g. tuberculosis) or failure (e.g. endocarditis) with inadequate treatment.

COMBINATIONS

In certain situations, combination therapy is indicated:

• to reduce the likelihood of drug resistance (e.g. tuberculosis)

• to provide synergy (e.g. benzylpenicillin and gentamicin for treating *Streptococcus viridans* and enterococcal endocarditis)

• where there is known or likely mixed infection (e.g. cefuroxime and metronidazole for gastrointestinal or biliary tract infections)

• to treat infection empirically where the likely organism is unknown (e.g. septic shock, neonatal sepsis, severe community-acquired pneumonia). In practice, this is the commonest reason

• where clinical studies have shown that survival is improved (e.g. piperacillin and gentamicin in Gram-negative infections in immunocompromised hosts)

• to cover the possibility of initial antibiotic resistance (e.g. vancomycin and cefotaxime where multiresistant *S. pneumoniae* is endemic).

Other forms of combination therapy include:

• a β-lactam antibiotic and β-lactamase inhibitor (e.g. amoxycillin and clavulanic acid)

• imipenem and cilastatin (a renal enzyme inhibitor reducing renal metabolism)

• probenecid with penicillins, reducing renal and cerebrospinal fluid (CSF) excretion, thereby increasing CSF and serum levels.

Combination therapy must be balanced by the potential disadvantages, namely interaction before administration (e.g. chloramphenicol and erythromycin), additional cost and combined adverse effects.

CHEMOPROPHYLAXIS

Prophylaxis can be broadly divided into primary (preventing initial infection or disease) and secondary (preventing recurrent disease), although these terms are used loosely. Examples of prophylaxis in medicine include:

- penicillin for preventing recurrent attacks of rheumatic fever
- amoxycillin (or other) for preventing endocarditis during dental and other operative procedures in patients with valvular lesions
- co-trimoxazole for preventing *Pneumocystis carinii* pneumonia in human immunodeficiency virus (HIV) patients with CD4 counts of $<200 \times 10^9$ cells/litre
- acyclovir for preventing herpes simplex and ganciclovir for preventing cytomegalovirus (CMV) infections after bone-marrow transplantation
- rifampicin and isoniazid for preventing subsequent relapse of a subclinical primary tuberculous infection
- rifampicin for preventing spread of *N. meningitidis* and *H. influenzae* nasopharyngeal carriage from a close contact of the index case
- erythromycin for preventing *Bordetella pertussis* infection in unvaccinated close home contacts.

Surgical antibiotic prophylaxis is mainly primary and should be given to the following categories of patients:

- those undergoing 'clean-contaminated' or 'contaminated' surgery (e.g. most gastrointestinal, biliary and gynaecological surgery)
- those receiving a prosthetic implant (e.g. orthopaedic and vascular)
- those with reduced resistance to infection
- where there is a break in aseptic technique intra-operatively.

The purpose of surgical prophylaxis is to cover the operative period when the site may become contaminated: antibiotics should not be given for more than 24 hours.

ANTIBIOTIC POLICY

The aim of an antibiotic policy is to provide a guide for the initial antibiotic therapy of a clinical syndrome. Supplementary to this is the achievement of an effective yet economic policy with reduced likelihood of encouraging the development of resistant organisms. Once the diagnosis has been confidently established and especially where an organism

has been recovered, the antibiotic therapy can be modified accordingly. To be effective, the policy must be:
• straightforward, sensible and applicable to the types of infection seen within the hospital
• cost conscious (but not cost driven) and advise antibiotics that are routinely stocked by the pharmacy and available in the emergency drugs cupboard
• well publicised, audited at regular intervals and safe.

To formulate the policy, a committee should be formed including a microbiologist, a pharmacist and a physician. This group would decide on clinical infections to be included, look at the organisms implicated, examine the local resistance patterns and choose the optimal antibiotic from an appropriate shortlist. Several factors have to be taken into account when writing the policy (e.g. acceptable failure rate, cost, available formulations), and thought must also be given to the age group of the hospital population, the likelihood of the infection being nosocomial, and the possibility of concomitant renal or liver compromise. The policy must be well publicised and supported by all grades of staff to be successful.

EMPIRICAL ANTIBIOTICS FOR SPECIFIC INFECTIONS

This section gives advice on the choice of antibiotic(s) when the microbiological cause is not known. The route of administration is IV unless stated otherwise.

Infection/age of patient	*Most likely organisms*	*Antimicrobials*
Central nervous system infections		
Meningitis		
Community acquired, age <1 month	'Coliforms', group B β-haemolytic *Streptococcus*, *Listeria*	Cefotaxime and ampicillin
Hospital acquired	As above but also *P. aeruginosa*	Ceftazidime and ampicillin
Age 1 month–5 years	*H. influenzae* (type b), *N. meningitidis*, *S. pneumoniae*	Cefotaxime
Age >5 years	*N. meningitidis*, *S. pneumoniae*	Cefotaxime (or benzylpenicillin or chloramphenicol if rash present)

Infection/age of patient	*Most likely organisms*	*Antimicrobials*
Intraventricular shunt or post-neurosurgery	*S. epidermidis*, proprionibacterium, *S. aureus*, 'coliforms', *P. aeruginosa*	Vancomycin and ceftazidime (consider intraventricular gentamicin and vancomycin)
Cerebral abscess	*Streptococcus milleri*, other streptococci, 'coliforms', anaerobes, *P. aeruginosa* (otitic source), *S. aureus* (systemic)	Benzylpenicillin, metronidazole and cefotaxime (ceftazidime if otitic)
Encephalitis	Herpes simplex, enteroviruses, influenza, *Mycoplasma pneumoniae*	Acyclovir (doxycycline if *M. pneumoniae* suspected)
Extradural spinal abscess	*S. aureus*, 'coliforms'	Flucloxacillin and cefotaxime
Bone and joint infections		
Osteomyelitis		
Age <5 years	*S. aureus*, *Streptococcus pyogenes*, *H. influenzae* (type b)	Flucloxacillin and cefotaxime
Age >5 years to adult	*S. aureus*	Flucloxacillin and fucidin (oral)
Vertebral	*S. aureus*, 'coliforms'	Flucloxacillin and cefotaxime
Complicating haemoglobinopathy	*S. aureus*, *Salmonella*	Flucloxacillin and ciprofloxacin
Complicating post-operative metal implant	*S. aureus*, *S. epidermidis*	Vancomycin
Septic arthritis		
Age <5 years	*S. aureus*, *S. pyogenes*, *H. influenzae* (type b)	Flucloxacillin and cefotaxime

Infection/age of patient	*Most likely organisms*	*Antimicrobials*
Age $>$5 years	*S. aureus*, *S. pyogenes*, group B β-haemolytic *Streptococcus*, *N. gonorrhoeae*, 'coliforms'	Flucloxacillin and cefotaxime
Prosthetic joint	*S. aureus*, *S. epidermidis*	Vancomycin
Gastrointestinal infections		
Cholecystitis, cholangitis or diverticulitis	Anaerobes, 'coliforms', enterococci	Cefuroxime and metronidazole, or co-amoxiclav, or gentamicin, ampicillin and metronidazole
Liver, pelvic or intra-abdominal abscess	*S. milleri*, other streptococci, anaerobes, enterococci, 'coliforms', *S. aureus*	Co-amoxiclav, or gentamicin, ampicillin and metronidazole
Diarrhoea or dysentery	*Campylobacter*, *Salmonella*, *Shigella*	Ciprofloxacin (antibiotics infrequently indicated; can usually be given orally)
Genitourinary tract infections		
Cystitis	'Coliforms', enterococci, *P. aeruginosa*, *Staphylococcus saprophyticus*, *S. aureus*	Trimethoprim (oral) or co-amoxiclav (oral)
Pyelonephritis	As above	Cefuroxime
Perinephric abscess	'Coliforms', *S. aureus*	Cefotaxime and flucloxacillin

Infection/age of patient	*Most likely organisms*	*Antimicrobials*
Pelvic inflammatory disease	*N. gonorrhoeae*, *C. trachomatis*, 'coliforms', anaerobes, enterococci	Co-amoxiclav (oral) and doxycycline (oral), or metronidazole (oral) and doxycycline (oral)
Respiratory tract infections		
Pneumonia		
Community acquired	*S. pneumoniae*, *S. aureus*, *H. influenzae*, *Coxiella*, *Mycoplasma*, *Legionella*, *C. psittaci*, *C. pneumoniae*	Cefuroxime and erythromycin
Hospital acquired	'Coliforms', *P. aeruginosa*, *S. aureus*, *Enterobacter*, *Acinetobacter*, *S. pneumoniae*, *Legionella*	Cefotaxime and ciprofloxacin
Aspiration	Anaerobes, streptococci, 'coliforms', *S. aureus*	Cefuroxime and metronidazole
Infective exacerbation of chronic lung disease	*S. pneumoniae*, *H. influenzae*, *M. catarrhalis*	Co-amoxiclav (oral), or co-trimoxazole (oral), or cefuroxime and erythromycin (if severe)
Ear, nose, throat and eye infections		
Otitis media, sinusitis		
Acute	*S. pneumoniae*, *S. pyogenes*, *H. influenzae*	Amoxycillin (oral), or co-trimoxazole (oral) or co-amoxiclav (oral)

Infection/age of patient	*Most likely organisms*	*Antimicrobials*
Chronic	As above, also anaerobes, 'coliforms'	Co-amoxiclav (oral)
Otitis externa	*S. aureus*, *P. aeruginosa*	Flucloxacillin (oral), topical agents
Tonsillitis	*S. pyogenes*	Benzylpenicillin or amoxycillin (oral)
Epiglottitis	*H. influenzae* (type b)	Cefotaxime
Dental abscess, gingivitis	Anaerobes, streptococci	Metronidazole (oral) or co-amoxiclav (oral)
Conjunctivitis	*S. aureus*, *S. pneumoniae*, *H. influenzae*	Chloramphenicol (topical)
Skin and subcutaneous tissue		
Cellulitis		
Normal host	*S. pyogenes*, *S. aureus*	Benzylpenicillin and flucloxacillin
Diabetic	As above, also anaerobes, enterococci, 'coliforms'	Co-amoxiclav, or cefotaxime, ampicillin and metronidazole (if severe)
Leg ulcer	*S. aureus*, anaerobes, *S. pyogenes*, group B β-haemolytic *Streptococcus*	Local cleansing flucloxacillin (oral)
Wound infections associated with		
Clean surgery	*S. aureus*	Flucloxacillin (oral)
Dirty surgery	Anaerobes, *S. aureus*, 'coliforms'	Co-amoxiclav (oral), or flucloxacillin (oral) and metronidazole (oral)

Infection/age of patient	*Most likely organisms*	*Antimicrobials*
Necrotising fasciitis	*S. pyogenes*, other streptococci, *S. aureus*, anaerobes, 'coliforms'	Benzylpenicillin, gentamicin and metronidazole
Bite	*S. aureus*, *Pasteurella multocida*, anaerobes	Co-amoxiclav (oral)
Cannula-related	*S. aureus*	Flucloxacillin (oral)
Septicaemia		
Origin/association		
Unknown	Gram-positive, Gram-negative, and/or anaerobes	Ampicillin, gentamicin and metronidazole
IV cannula	*S. aureus*, *S. epidermidis*	Vancomycin
Urinary tract	'Coliforms', *P. aeruginosa*	Cefuroxime or ciprofloxacin
Biliary tract	'Coliforms', anaerobes, enterococci	Cefuroxime and metronidazole, or co-amoxiclav
Skin, subcutaneous tissue	*S. aureus*, *S. pyogenes*	Benzylpenicillin and flucloxacillin
Neutropenia	'Coliforms', *P. aeruginosa*	Gentamicin and piperacillin
Neonatal	'Coliforms', *Listeria*, group B β-haemolytic *Streptococcus*	Gentamicin and ampicillin

CHAPTER 4

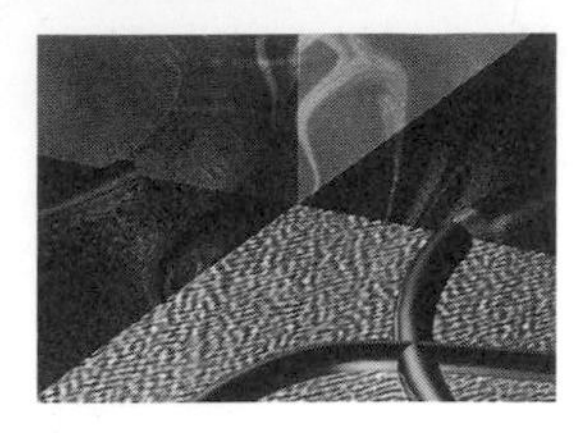

Hospital-Acquired Infection

Hospital-acquired (nosocomial) infections have been important for as long as infection has been understood. Their prevalence is 10 per cent, their incidence 5 per cent; they cost lives and money. Urinary tract, wound and respiratory tract are the commonest types. Nosocomial urinary tract infections (UTI) result in bacteraemia in 1–3 per cent, of which one in eight results in death. The current estimated annual cost is over $2 billion in the USA and £180 million in the UK. The causative organisms come from three main sources:

- the hospital environment and equipment (e.g. *Legionella pneumophila*, *Clostridium difficile*)
- the patient's own microbial flora (e.g. *Staphylococcus aureus* wound infection)
- directly from other patients or staff, or indirectly (usually via the hands of staff) from other persons (e.g. multiresistant Gram-negative bacilli).

Infection can be transmitted by:

- aerosol (e.g. Legionnaire's disease, varicella zoster)
- faecal–oral (e.g. *C. difficile*, *Salmonella*)
- hand or body contact (e.g. multiresistant Gram-negative bacilli, MRSA)
- contaminated vehicles (e.g. food (*Clostridium perfringens*), equipment (hepatitis B))
- blood products (e.g. hepatitis B, hepatitis C).

ORGANISATION OF INFECTION CONTROL

Control of hospital infection depends upon adherence to basic infection control principles and the existence of several factors, including an infection control team, regular surveillance of high-risk units (e.g. intensive care), monitoring of laboratory results, and the development of policies on topics such as isolation, disinfection as well as antibiotic usage.

PRINCIPLES OF ISOLATION

The level of isolation will depend on the micro-organism, the site of infection, the susceptibility of others and the other medical needs of the patient, which may override the need for strict isolation. With this in mind, the following principles should be observed:

- isolation using a negative-pressure ventilation system is usually required when the patient poses a risk to other patients and staff
- disposable plastic aprons should be used for all intimate patient care when soiling of clothing is likely, when dealing with body sites where there is a break in the normal defence system (e.g. wound and catheter care), when the patient has a skin disorder and is a likely heavy disperser of organisms, and when the patient is under protective isolation
- handwashing should be carried out after contact with every patient. Health-care workers should have intact skin and should cover cuts with a waterproof dressing. Soap and water is nearly always sufficient for hand cleansing but in high dependency areas such as intensive care, an antiseptic preparation is necessary
- gloves should be worn when there is possible contact with secretions, excretions or blood, or where gross hand contamination is likely
- masks, full-length gowns, overshoes and hats are only needed when dealing with a patient who is a heavy disperser of multiresistant organisms (e.g. MRSA) or when a transmissible viral haemorrhagic fever is suspected (e.g. Lassa fever)
- staff members who are unwell with symptoms consistent with infection should not come to work or should be deployed on non-clinical duties.

MAJOR ISSUES IN INFECTION CONTROL

Nosocomial infections result from micro-organisms which originate from an environmental or human source and cause disease at particular site(s), usually because of some hospital-related interference with the normal defence barriers. Issues relate to the type of micro-organism (multiresistant or easily transmissible) and the clinical site of disease.

MULTIRESISTANT INFECTIONS

These can be innately resistant (e.g. *Xanthomonas maltophilia*) or selected resistant strains through hospital and patient antibiotic use (e.g. multiresistant *Klebsiella aerogenes*). The most important are:

- multiresistant Gram-negative bacilli (e.g. *Pseudomonas*, *Acinetobacter*)
- methicillin-resistant *S. aureus* (MRSA)
- vancomycin-resistant enterococci
- multiresistant pneumococci (see Chapter 7)

• multidrug-resistant *Mycobacterium tuberculosis* (MDRTB) (see Chapter 7).

MRSA are becoming an increasing problem and few hospitals are spared. They show the following characteristics:

• resistance to all β-lactam antibiotics and only reliably sensitive to vancomycin

• containable by a strict policy of isolation and regular screening of the index patient and contacts (staff and other patients). Elimination can usually be achieved by eradication of patient carriage (e.g. mupirocin nasal ointment for nasal carriage) and treatment of any infection (e.g. vancomycin and rifampicin therapy)

• particularly problematic in patients with skin disorders who are heavy shedders of colonised skin scales and often the origin of outbreaks.

Vancomycin-resistant enterococci (usually *Enterococcus faecium*) show the following characteristics:

• they are increasingly recognised as causing significant disease in debilitated hosts

• they are usually also resistant to ampicillin and the aminoglycosides (agents normally active against enterococci)

• they are being recovered from patients in large health-care centres, where cephalosporin and vancomycin use is widespread.

RARE, BUT TRANSMISSIBLE AND POTENTIALLY SEVERE, INFECTIONS

Several classic infectious diseases which exacted a high mortality in the pre-antibiotic and pre-immunisation era are still occasionally observed. In addition, certain imported conditions have the potential for person-to-person spread. The major infections in this group are:

• diphtheria

• Lassa, Marburg, Ebola and Congo–Crimean haemorrhagic fevers

• rabies, pulmonary anthrax, pulmonary plague.

INFECTIONS ACQUIRED FROM THE HOSPITAL ENVIRONMENT OR EQUIPMENT

Multiresistant Gram-negative bacilli, *Legionella* species and *C. difficile* are the most important in this category. Legionellosis has caused large outbreaks of hospital-acquired pneumonia when it has contaminated the ventilation systems, and sporadic cases have been linked to hospital showers. Control policies now exist in hospitals to reduce the risk of such outbreaks occurring and emphasise the need for control of hot-water temperature, cleaning and disinfection of water tanks and humidifiers and avoidance of water-cooled air-conditioning systems.

Multiresistant Gram-negative bacilli mainly result from in-hospital selection pressure through intensive medical and surgical practice and broad-spectrum antibiotic use, and are unavoidable. They survive well in the environment and are readily transmitted between patients. Cross-infection to unaffected patients can be restricted through isolation, hand-washing and standard infection control nursing procedures.

Clostridium difficile has become a common and important nosocomial infection. Its rise has paralleled the equally swift increase in broad-spectrum antibiotic use which selects out *C. difficile* by interfering with the commensal gastrointestinal flora (colonisation resistance). The following procedures should be observed:

- individual cases should be isolated, the need for antibiotics reviewed, and treatment with metronidazole or vancomycin instituted
- key areas in controlling an outbreak are immediate and effective isolation of all patients with diarrhoea, urgent sample collection and same-day toxin testing, immediate and rigorous cleaning of the patients' adjacent environment, and review of the antibiotic policy to reduce broad-spectrum antibiotic recommendations.

OUTBREAK INFECTIONS

An outbreak is defined as two or more cases that are linked in time or place. They tend to fall into three types, the 'point source' in which there is a single exposure that is not repeated (e.g. *C. perfringens* food poisoning), the 'continuous source' where the source remains infective (e.g. a surgeon with hepatitis B) and the 'propagating source' where the original source has resulted in secondary spread (e.g. *Salmonella* infection). The major organisms that cause hospital outbreaks are:

- enteric pathogens (*Salmonella, Shigella, C. difficile*)
- food poisoning pathogens (*C. perfringens, S. aureus*)
- respiratory pathogens (*Legionella* spp., influenza)
- skin and subcutaneous tissue pathogens (*S. aureus*)
- parenteral pathogens from contaminated intravenous (IV) preparations (e.g. parenteral feeds) or an infectious carrier (hepatitis B).

CLINICAL SITE OF DISEASE

The commonest sites of nosocomial infection are the urinary tract, respiratory tract, and skin and subcutaneous tissues. The majority of disease at these sites is innocuous but occasionally may be life-threatening (e.g. septicaemia complicating UTI). Numerically less, but individually more important are infections of prosthetic heart valves, orthopaedic implants, intraventricular shunts, and intravascular devices and grafts, all of which may be hospital-acquired.

URINARY TRACT

Hospital-acquired UTI is characterized by the following:

- it is the commonest nosocomial infection, accounting for up to one-third of all infections acquired in hospital and developing in 2–5 per cent of all admissions. Eighty per cent of these are associated with catheterisation, a procedure which is carried out in up to 15 per cent of patients during their stay
- it is complicated by bacteraemia in 1–3 per cent of patients, of whom one in eight die as a direct result.

Following catheterisation, the incidence of bacteriuria using closed drainage increases by 5–10 per cent per day and by 30 days it is almost universal. Organisms ascend urethrally, external to the catheter; initial asymptomatic bacteriuria progresses in many patients to become a symptomatic UTI. In long-term catheterised patients, up to two-thirds of febrile episodes are as a result of urinary infection. The organisms involved are typically 'coliforms' (*Escherichia coli*, *Proteus* spp., etc.) and enterococci; occasionally, *P. aeruginosa* and *Candida* are responsible. Many cause little illness and disappear when the catheter is removed. However, there is a risk of Gram-negative septicaemia at any time, especially if a patient with bacteriuria is re-catheterised or undergoes surgical procedures on the urinary tract or prostate gland.

SKIN AND SUBCUTANEOUS TISSUE

Post-operative wound infections may result from intra-operative contamination with endogenous flora, post-operative inoculation of endogenous flora (e.g. *S. aureus* from nasal carriage), or exogenous contamination (environmental or cross-infection). The most important of these is intra-operative contamination and the degree of contamination of the operation site is the main factor in the classification of the risk for post-operative infection.

- 'Clean' operations—minimal risk (no infection encountered, no break in aseptic technique, no colonised viscus opened). An example is hernia repair
- 'Clean-contaminated' operations—moderate risk (minimal contamination from opened colonised viscus or break in aseptic technique). An example is uncomplicated appendectomy
- 'Contaminated' operations—significant contamination from colonised opened viscus, acute inflammation without suppuration, or traumatic wounds within 4 hours of injury. An example is major colorectal surgery
- 'Dirty' operations—frank pus present, perforated viscus or traumatic wound more than 4 hours after injury. An example is perforated colonic diverticulum with paracolic abscess.

The organisms recovered reflect the colonisation flora of the viscus involved. For bowel surgery, 'coliforms' and *Bacteroides fragilis* are most frequently isolated. Prophylactic antibiotics are dealt with in Chapter 3. Host (age, obesity, immunosuppressives, underlying major illness) and surgical (technical skill) factors also play an important part in determining the risk of post-operative wound infection.

RESPIRATORY TRACT INFECTIONS

The incidence of nosocomial pneumonia is between four and eight per 1000 admissions to hospital. It is higher in comatosed patients, those with left ventricular failure and those anaesthetised and intubated; it is as high as 41 per cent in ventilated patients, increasing by 1 per cent for every day of ventilation. The organism is likely to be Gram-negative, with 'coliforms', *P. aeruginosa* and *Acinetobacter* predominating; enterococci (especially if on broad-spectrum cephalosporins) and *S. aureus* infections are also seen. *Legionella pneumophila* is dealt with above.

FOREIGN BODY-RELATED INFECTIONS

These are more serious nosocomial infections, often aided by the presence of a foreign body (e.g. prosthetic joint), when a much lower level of contamination is necessary to cause an infection. Infections are characteristically chronic and due to low-grade pathogens (e.g. *S. epidermidis*) where production of a glycocalyx enables adherence to the implant. Distinction from contamination is always a problem: recovery usually necessitates removal of the prosthesis.

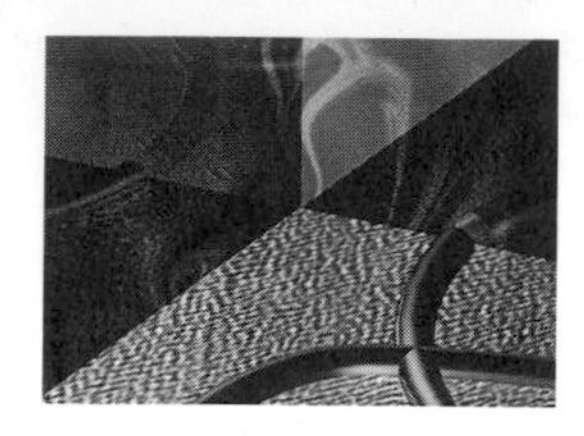

Pyrexia of Unknown Origin

Pyrexia of unknown origin (PUO) is a rather imprecise term which should be reserved for prolonged fever, the cause of which has not been established even after proper history-taking, detailed physical examinations and an initial phase of investigations over a period of 2 weeks or so, preferably in a hospital setting.

LIKELY CAUSES

The majority of the common viral infections which often present just with fever will have settled by the end of this period. Other infections which initially present with a fever prodrome will have developed specific diagnostic features by this time, e.g. hepatitis, rash-producing illnesses. The diagnosis of malaria, typhoid, brucellosis, septicaemias, clinically silent pneumonias, most pulmonary and haematological malignancies, should become evident during the initial phase of investigations. In patients who remain febrile beyond this stage, the eventual diagnosis will generally be from one of the following categories.

INFECTION

This still accounts for a significant proportion of PUO cases, but the longer the duration of fever, the less likely becomes an infective aetiology.

Tuberculosis (particularly in patients of Indian subcontinental origin): may cause diagnostic problems in the following clinical types: cryptic or disseminated (miliary opacities not yet in chest X-ray), mediastinal lymphadenitis (may not be apparent in routine chest X-ray), abdominal tuberculosis (fever and mild abdominal pain may be the only features)

Hidden abscesses: these are commonly intra-abdominal, i.e. subphrenic, intrahepatic, pancreatic, renal or perirenal, splenic, paracolic or pelvic

Endocarditis: significance of a seemingly innocent systolic murmur may have been missed and blood culture is negative, either because of prior antibiotics or because the patient has Q fever.

Other infective causes

Glandular fever: throat involvement may be minimal and Paul–Bunnell test may be negative initially

Cytomegalovirus (CMV) and Coxiella burnetii: may cause subacute granulomatous hepatitis, presenting only with fever and low-grade liver function abnormalities

Tick typhus: may be missed if adequate travel history is not taken and the significance of an eschar overlooked

Brucellosis, toxoplasmosis, amoebic liver infection, visceral leishmaniasis and trypanosomiasis: can present as prolonged fever in appropriate circumstances

Chronic prostatitis, urinary tract infections and pelvic sepsis: important causes of recurrent fever.

Fever in the immuncompromised

In neutropenic cancer patients, fever is usually bacterial in aetiology, but blood cultures are often sterile. Fungal infections such as *Candida*, *Mucor* and *Aspergillus* may also be the cause.

In cell-mediated immunity (CMI)-deficient human immunodeficiency virus (HIV)-infected patients, fever may be a feature of HIV infection itself. *Pneumocystis carinii* pneumonia, cryptococcal infections, cerebral toxoplasmosis, CMV disease, tuberculosis and *Mycobacterium avium* complex (MAC) infection all may present only as fever.

Fever due to self-induced infection

Fever arising from injection of contaminated fluids, e.g. sputum, urine, water, etc. The polymicrobial nature of these infections is often a clue to the diagnosis.

NEOPLASMS

Hodgkin's disease and other lymphoreticular malignancies are the commonest neoplastic causes of PUO, followed by hypernephroma and metastatic carcinoma. Intra-abdominal malignances are particularly liable to present as PUO.

CONNECTIVE TISSUE DISEASES AND MISCELLANEOUS CONDITIONS

Fever may be the only manifestation in the different forms of connective tissue diseases. Other conditions which may present as PUO are: sarcoidosis, multiple pulmonary embolism, Crohn's disease, Whipple's disease and familial Mediterranean fever.

DIAGNOSTIC APPROACH

Before considering further laboratory tests, a thorough reappraisal of the situation is essential, including travel history, epidemiological background, occupational history and thorough clinical examination. Results of the investigations available should be reviewed and any abnormality, however minimal, should be noted, as this may indicate the direction of further diagnostic approach. Temptation to try 'blind' antibiotic therapy should be resisted unless the particular situation in a dangerously ill patient warrants such an action. The following further investigations are often helpful.

SEROLOGICAL TESTS

As the patient has already been ill for 2 weeks or more, the second sample of serum for antibodies should now be taken and tested, particularly for CMV, *Coxiella burnetii*, *Chlamydia* and *Mycoplasma*. Tests for *Toxoplasma*, *Amoeba*, *Brucella* and *Rickettsia* should be performed in appropriate circumstances if not already done. Paul–Bunnell screening test should be repeated.

SEARCH FOR INFECTIVE AGENTS

Further blood cultures (in the case of prior antibiotic therapy), culture of induced sputum or early morning gastric aspirate and urine (in suspected disseminated tuberculosis) may help. In immunocompromised patients the following tests are useful: CMV early antigen in blood; *P. carinii*, MAC, *M. tuberculosis* in induced sputum or broncho-alveolar lavage fluid; MAC in bone-marrow culture and cryptococcal antigen in serum.

ULTRASOUND, ISOTOPE AND COMPUTERISED TOMOGRAPHY SCANS

These are the most valuable investigations at this stage, particularly if an intra-abdominal pathology is suspected. Combined liver and lung isotope scanning is useful for demonstrating subdiaphragmatic abscess. Gallium-67 scans have limited usefulness in identifying intra-abdominal abscesses, and scanning with indium-111 labelled leucocytes is more useful in such a situation. Isotope scans are also helpful in locating bone lesions. Computerised tomography scans are particularly helpful in demonstrating retroperitoneal pathology and identifying intrathoracic tumours/

pathologies and miliary lung lesions not well visualised in ordinary chest X-ray.

BIOPSY

Liver biopsy is useful if there is biochemical evidence of hepatic dysfunction, even if this takes the form of isolated elevation of alkaline phosphatase or gamma GT. It may establish the diagnosis of tuberculosis, sarcoidosis, disseminated lymphoma or non-caseating hepatic granulomatosis (e.g. due to Q fever, CMV, *Brucella*). Any enlarged gland or skin lesion should be biopsied. Peritoneal biopsy is a valuable procedure for suspected tuberculous peritonitis. Scalene node biopsy may establish the diagnosis of sarcoidosis in the absence of other glandular enlargement. Mediastinoscopic biopsy of enlarged mediastinal glands may show tubercular, sarcoidal or lymphomatous pathology. Bone-marrow biopsy may be helpful in disseminated tuberculosis, kala-azar and various infiltrative or neoplastic disorders. Intestinal biopsy may show Whipple's or Crohn's disease. Percutaneous biopsy of a deep-seated lesion identified by scanning technique is used increasingly.

THERAPEUTIC TRIAL

If tuberculosis is strongly suspected, a trial of anti-tuberculous drugs may be justified when other methods to achieve a diagnosis have failed and culture results are awaited.

Occasionally, even after exhaustive investigations, the diagnosis is still not apparent. In some cases, the fever settles spontaneously without a diagnosis ever being achieved; in others, the nature of the disease becomes apparent only with the passage of time. Finally, the possibility that the patient might be faking fever should be considered if the general condition remains good or if the pulse rate fails to rise during fever.

SECTION 2

System-Based Infections

CHAPTER 6

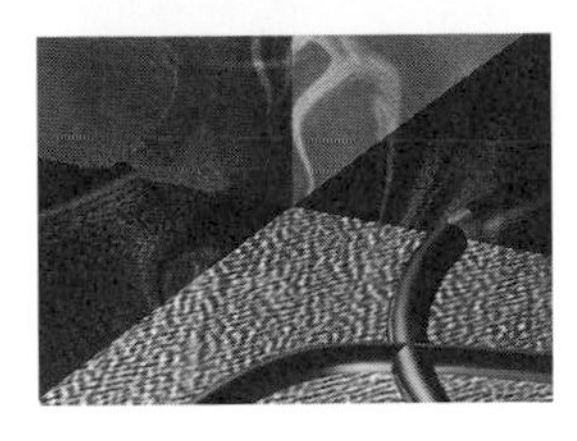

Eye and Upper Respiratory Tract Infections

CLINICAL SYNDROMES

Conjunctivitis, keratitis and uveitis

Major presenting feature: *painful red eye.*

Causes

Conjunctivitis

- ADENOVIRUS
- *CHLAMYDIA TRACHOMATIS*
- *Streptococcus pneumoniae*, *Neisseria gonorrhoeae*, *Staphylococcus aureus*, *Haemophilus influenzae*
- Enteroviruses, herpes simplex, varicella zoster, measles, rubella, *Mycobacterium tuberculosis*, *Neisseria meningitidis*, leptospirosis, Stevens–Johnson syndrome.

Keratitis

- HERPES SIMPLEX
- *Staphylococcus aureus*, *S. pneumoniae*, *Streptococcus pyogenes*, *Pseudomonas aeruginosa*, *N. gonorrhoeae*
- Varicella zoster, measles, adenovirus, mumps
- *Fusarium*, acanthamoeba, *Treponema pallidum*, *Mycobacterium tuberculosis*, *Mycobacterium leprae*.

Uveitis

- Herpes simplex, varicella zoster
- *Treponema pallidum*, *N. gonorrhoeae*, brucellosis, toxoplasmosis, *M. tuberculosis*, *M. leprae*, Lyme disease, typhoid, leptospirosis.

Distinguishing features

	Keratitis	*Conjunctivitis*	*Uveitis*
Visual acuity	impaired	normal	impaired
Pain	severe	gritty	severe
Discharge	only if bacterial	present	absent
Photophobia	common	uncommon	common
Blepharospasm	common	uncommon	rare
Ciliary injection	present	absent	present
Cornea	ulcer/oedema	clear	clear
Pupil	normal	normal	normal/small, irregular
Pupillary reaction	normal	normal	sluggish

Complications

Conjunctivitis: ophthalmia neonatorum, keratitis, uveitis

Keratitis: corneal scarring, ulceration and perforation (leading to blindess), hypopyon, uveitis

Uveitis: hypopyon, synechiae, raised intraocular pressure.

Investigations

- Full blood count (FBCt), differential white cell count (WCCt) and erythrocyte sedimentation rate (ESR)
- Chest X-ray (CXR) (for evidence of hilar lymphadenopathy indicating tuberculosis or sarcoidosis)
- Conjunctival/corneal scrapings for cytology
- Swabs for bacterial, viral, chlamydial and fungal culture
- Acute and convalescent serology for infectious agents (as above).

Differential diagnoses

- Acute glaucoma, subconjunctival haemorrhage, foreign body, trauma, episcleritis, scleritis.

Treatment

Conjunctivitis

Bacterial: topical antibiotics (e.g. chloramphenicol), usually sufficient for bacterial infection

Chlamydial: topical tetracycline or erythromycin, and oral co-trimoxazole, erythromycin or tetracycline.

Keratitis

Bacterial: subconjunctival and parenteral drugs usually needed in addition to intensive topical antibiotics

Herpetic: oral acyclovir.

Uveitis

Herpetic: oral acyclovir, topical corticosteroid and mydriatic.

Prevention

- Erythromycin and tetracycline are effective in preventing ophthalmia neonatorum due to *N. gonorrhoeae* and *C. trachomatis*: silver nitrate protects only against *N. gonorrhoeae*. These agents can be used where there is high endemicity for either infection
- Children born to mothers with known gonococcal infection should be treated with parenteral antibiotics (e.g. benzylpenicillin).

Tonsillitis/pharyngitis

Major presenting feature: *sore throat.*

Causes

- *STREPTOCOCCUS PYOGENES*, groups C and G β-haemolytic streptococci
- EPSTEIN–BARR VIRUS (EBV)
- *CORYNEBACTERIUM DIPHTHERIAE*
- *Mycoplasma pneumoniae*, *Corynebacterium haemolyticum*, *Corynebacterium ulcerans*, *C. trachomatis*, fusobacterium, *N. gonorrhoeae*, *N. meningitidis*
- Adenovirus, coxsackie, herpes simplex, parainfluenza, cytomegalovirus (CMV), human immunodeficiency virus (HIV)
- *Toxoplasma*.

Distinguishing features

	S. pyogenes	*Glandular fever*	*C. diphtheria*
Age group	children	teenagers	children and young adults
Spread	droplets	saliva exchange	droplets
IP	2–5 days	14–42 days	2–5 days

	S. pyogenes	*Glandular fever*	*C. diphtheria*
Onset	abrupt	slow	abrupt
Tonsillar exudate	white, thin	white, thin	grey-green, thick, with possible spread to palate, pharynx or larynx
Palate	normal	petechiae	membrane (as above)
Lymph nodes	tender, no oedema	non-tender, no oedema	non-tender, oedema (bullneck)
Fever	high	high	low grade
Toxicity	moderate	moderate	marked
WCCt	raised, neutrophilic	raised, lymphocytic	raised, neutrophilic
Heterophile antibody	negative	positive	negative
Throat swab culture	positive	negative	positive

Complications

Streptococcus pyogenes: scarlet fever, quinsy, rheumatic fever, glomerulonephritis, otitis media, erythema nodosum, erythema multiforme, Henoch–Schönlein purpura

Epstein–Barr virus: haemolytic anaemia, thrombocytopenia, airways obstruction (tonsillar), splenic rupture, hepatitis, Guillain–Barré syndrome, rash with antibiotics, meningoencephalitis

Corynebacterium diphtheria: myocarditis, neuropathy (palate (3 weeks), ocular (4 weeks), respiratory (7 weeks), peripheral (10 weeks)), airways obstruction (laryngeal), glomerulonephritis.

Investigations

- FBCt with differential WCCt
- Heterophile antibody test (Paul–Bunnell or Monospot)
- Throat swab for *S. pyogenes*: if negative, acute and convalescent sera for antistreptolysin O (ASO) titre
- If diphtheria possible: throat swab for *C. diphtheria* (telephone laboratory); electrocardiogram (ECG) to detect myocarditis
- Throat swab for viral culture
- Blood culture (for *Fusobacterium necrophorum*)
- CXR for lung abscesses (*F. necrophorum*).

Differential diagnoses

If tonsillar or nearby swelling: quinsy, lymphoma, leukaemia, tuberculosis, cancer, retropharyngeal abscess.

Treatment

Streptococcus pyogenes

- Intravenous (IV) benzylpenicillin, oral amoxycillin or penicillin V
- Local analgesia (benzydamine or aspirin gargle).

Epstein–Barr virus

- Local analgesia (benzydamine or aspirin gargle)
- Consider IV hydrocortisone if upper airways obstruction or toxicity.

Corynebacterium diphtheria

- Intramuscular (IM)/IV antitoxin
- IV erythromycin or benzylpenicillin
- Bed rest, cardiac monitor.

Prevention

Streptococcus pyogenes

- Penicillin V for patients with a history of rheumatic fever.

Corynebacterium diphtheria

- Immunisation at 2, 3 and 4 months and at 4 years
- Treatment of carriers with oral erythromycin
- Post-exposure erythromycin prophylaxis for close contacts.

Parotitis and cervical lymphadenitis

Major presenting feature: *swelling and pain around the angle of the jaw.*

Causes

Parotitis

- MUMPS
- *STAPHYLOCOCCUS AUREUS*
- Parainfluenzae 3, coxsackie, CMV, *S. pyogenes*.

Cervical lymphadenitis

- *STREPTOCOCCUS PYOGENES*
- *STAPHYLOCOCCUS AUREUS*
- EBV, toxoplasmosis, HIV, *Eikenella corrodens*, *C. diphtheriae*, anaerobes, cat scratch disease, *Bacillus anthracis*, atypical mycobacteria, *M. tuberculosis*, CMV, Kawasaki disease.

Distinguishing features

	Mumps	*Suppurative parotitis*	*Cervical lymphadenitis*
Incidence	decreasing	uncommon	common
Age group	children	all ages	children
Spread	droplet or saliva	primary oral sepsis	primary oral or skin sepsis
IP	17–19 days	2–4 days	2–4 days
Onset	subacute	acute	acute
Parotid duct	red	exuding pus on massage	normal
Angle of the jaw	obscured	obscured	palpable
Tenderness	moderate	marked	marked
Overlying skin	normal	red	red/normal
WCCt	normal/low, lymphocytic	raised, neutrophilic	raised, neutrophilic
Amylase	raised	normal	normal

Complications

Mumps: meningitis, epididymo-orchitis, encephalitis, pancreatitis, oophoritis

Suppurative parotitis: septicaemia, massive neck oedema, respiratory obstruction, osteomyelitis

Cervical lymphadenitis: suppuration, septicaemia, jugular venous and intracranial sinus thromboses.

Investigations

- FBCt with differential WCCt
- Amylase
- Throat swab for bacterial and viral culture
- Parotid duct swab for bacterial culture
- Blood culture
- CXR (to look for hilar lymphadenopathy)
- Serology: mumps S and V antigens; viral screen; ASO titre.

Differential diagnoses

Parotitis

- Lymphoma, leukaemia, sarcoidosis, alcoholic liver disease, sialadenitis, salivary tumour, Sjögren syndrome, post-operative swelling, branchial cyst.

Cervical lymphadenitis

- Lymphoma, leukaemia, cancer, sarcoidosis.

Treatment

Mumps

- Symptomatic; analgesics, mouth washes.

Suppurative parotitis

- IV benzylpenicillin and flucloxacillin
- Surgical drainage may be needed.

Cervical lymphadenitis

- IV benzylpenicillin and flucloxacillin
- Surgical drainage may be needed if suppuration develops.

Prevention

Mumps

- Immunisation with live mumps vaccine as part of measles, mumps and rubella (MMR) administration.

Epiglottitis and laryngotracheobronchitis

Major presenting feature: *difficulty in breathing.*

Causes

Epiglottitis

- HAEMOPHILUS INFLUENZAE (TYPE B)
- *Staphylococcus aureus*, non-B capsulate *H. influenzae*, *S. pneumoniae*, *S. pyogenes*, varicella zoster.

Laryngotracheobronchitis (croup)

- PARAINFLUENZAE I
- Parainfluenzae 2 or 3, measles, respiratory syncytial virus (RSV), influenza, adenovirus, rhinovirus, *C. diphtheria*, *M. pneumoniae*.

Distinguishing features

	Viral croup	*Epiglottitis*
Incidence	common	rare
Season	winter	all year
Age group	<3 years	2–5 years

	Viral croup	*Epiglottitis*
History	days	hours
General condition	agitated	lethargic/toxic
Fever	low grade	high
Toxaemia	absent	present
Cough	brassy	weak/absent
Audible stridor	marked	moderate
Drooling saliva	absent	present
WCCt	normal/low, normal/lymphocytic	high, neutrophilic
X-ray of neck	subglottic swelling	epiglottic swelling

Complications

Epiglottitis: airway obstruction, hypoxia, septicaemic shock, vagally induced cardiopulmonary arrest

Laryngotracheobronchitis: respiratory failure, pulmonary oedema, pneumothorax, aspiration pneumonia.

Investigations

An anaesthetist must be called immediately if epiglottitis is suspected. No investigations should be performed until the personnel and equipment to secure the airway are available.

- FBCt with differential WCCt
- Blood culture (septicaemia common with epiglottitis)
- Throat swab for bacterial and viral culture
- Epiglottal culture (*only* at intubation)
- Nasopharyngeal aspirate for rapid viral antigen detection (RSV, parainfluenzae 1–3, adenovirus)
- If diphtheria possible (immigrant, throat features): throat swab for *C. diphtheria* (telephone laboratory); ECG
- Capillary/arterial gases or oxygen saturation
- Chest X-ray
- Lateral neck X-ray.

Differential diagnoses

- Foreign body, allergic reaction, retropharyngeal abscess, suppurative tracheitis, laryngomalacia, recurrent laryngeal nerve paralysis.

Treatment

Epiglottitis

- Secure airway (endotracheal tube or tracheostomy)

- Intensive care monitoring
- Cefotaxime (because *H. influenzae* type b resistance to ampicillin is 30 per cent and to chloramphenicol is 1–3 per cent neither can be used confidently).

Laryngotracheobronchitis
- Supplemental oxygen
- Humidification
- Corticosteroids should be used in severe cases.

Prevention

Epiglottitis
- Rifampicin prophylaxis for all family contacts where there is a child <5 years
- *Haemophilus influenzae* type b (Hib) vaccine at 2, 3 and 4 months.

Bacterial sinusitis

Major presenting features: *facial pain, blocked nose, fever, post-nasal drip.*

Causes

Predisposing:
- viral infection, cystic fibrosis, Kartagener's syndrome (immotile cilia), nasal polyps, allergic rhinitis, foreign body, septal deviation, immune deficiency, dental abscess, nasopharyngeal tumours, nasal packing.

Bacterial – acute:
- *Haemophilus influenzae, S. pneumoniae, S. pyogenes, Moraxella catarrhalis, S. aureus.*

Bacterial – chronic:
- Anaerobes, *Streptococcus milleri*, 'coliforms', *P. aeruginosa*
- *Haemophilus influenzae, S. pneumoniae, S. pyogenes, M. catarrhalis, S. aureus*
- *Mycobacterium tuberculosis*, fungi.

Complications

- Chronic sinusitis, meningitis, intracranial abscess, subdural empyema, osteomyelitis.

Investigations

- Sinus X-ray (for mucosal thickening or fluid level)
- Imaging with ultrasound scan (USS) or computerised tomography (CT) scan (usually reserved for complicated disease)

- Antral lavage (for chronic disease): specimens for microscopy, Gram and Ziehl–Nielsen stain; bacterial, mycobacterial and fungal culture
- Antroscopy and biopsy (for chronic disease).

Differential diagnoses

- Chronic hyperplastic allergic sinusitis, sarcoidosis, malignancy, upper molar dental abscess.

Treatment

- Antibiotics:
  - acute sinusitis: amoxycillin, co-trimoxazole or co-amoxiclav
  - chronic sinusitis: co-amoxiclav
- Decongestants
- Surgery (for chronic disease): antral lavage or antrostomy (simple and radical)
- Topical corticosteroids (if allergic component).

Prevention

- Surgery (for polyps, deviated nasal septum).

Bacterial otitis media

Major presenting features: *earache*, *fever*, *ear discharge*, *deafness*.

Causes

Predisposing:

- viral infection, enlarged adenoids, immune deficiency, nasopharyngeal tumours.

Bacterial – acute:

- *Haemophilus influenzae*, *S. pneumoniae*, *S. pyogenes*, *M. catarrhalis*, *S. aureus*
- *Mycoplasma pneumoniae*, *C. trachomatis*.

Bacterial – chronic:

- anaerobes, 'coliforms', *S. milleri*, *P. aeruginosa*
- *Haemophilus influenzae*, *S. pneumoniae*, *S. pyogenes*, *M. catarrhalis*, *S. aureus*
- *Mycobacterium tuberculosis*, fungi.

Complications

- 'Glue ear' (persistent middle ear effusion), mastoiditis, labyrinthitis, meningitis, intracranial abscess, facial nerve paralysis, chronic perforation of the drum, cholesteatoma.

Investigations

- Ear swab for bacterial and fungal culture if chronic perforation
- Throat swab for *S. pyogenes*.

Differential diagnoses

- Viral infection, otitis externa, foreign body, 'glue ear'.

Treatment

- Paracetamol suspension
- Antibiotics:
  - acute otitis media: amoxycillin, co-trimoxazole or co-amoxiclav
  - chronic otitis media: co-amoxiclav
- For complicating 'glue ear':
  - myringotomy
  - grommet insertion.

Prevention

- Adenoidectomy (indicated for children with prolonged glue ear).

SPECIFIC INFECTIONS

Trachoma and inclusion conjunctivitis

Epidemiology

Trachoma

Trachoma is caused by *Chlamydia trachomatis*, and is the commonest infective cause of blindness in the world. It is endemic in the hot, dry areas of the tropics and subtropics.

- Transmission is from an infected eye by hands, materials and flies
- Young children are most at risk
- It is associated with inadequate public and personal hygiene
- Repeated infections occur
- The incubation period (IP) is 5–7 days.

Inclusion keratoconjunctivitis

Inclusion keratoconjunctivitis is common in temperate climates and is also caused by *C. trachomatis*.

- Transmission is usually sexual (adults), vertical (ophthalmia neonatorum) or by direct or indirect contact (children)

- There is often associated chlamydial genital infection in the patient or partner
- The IP is 5–12 days.

Pathology and pathogenesis

Chlamydia are intracellular pathogens dependent on the host cell for energy. They have two major forms, the infectious elementary body and the non-infective reticulate body. *Chlamydia trachomatis* is divided into 15 serovars; of these:

- A, B and C cause trachoma
- D, E, F and G cause genital infections, ophthalmia neonatorum, and adult ocular infections in developed nations
- L1, L2 and L3 cause lymphogranuloma venereum.

In trachoma an acute inflammatory response with purulent conjunctivitis and follicular reaction in the superior tarsal conjunctiva follows infection. Fibrous tissue and new blood vessels (pannus) form with repeated infections, leading to blindness. The eyelids become thickened and everted, leaving the conjunctivae susceptible to damage from infection and dust. In inclusion conjunctivitis, follicles are more pronounced on the lower tarsal conjunctiva and scarring is rare.

Clinical features

- Acute purulent follicular conjunctivitis
- Lachrymal gland and pre-auricular gland enlargement.

Complications

- Recurrent bacterial secondary infections
- Corneal scarring, new vessel formation
- Lid eversion
- Blindness.

Diagnosis

Diagnosis can be achieved by looking for antigen or inclusion bodies on conjunctival smears or by culture. Serology is unhelpful, but measurement of tear antibody may be useful.

Treatment

For the acute attack, treatment with tetracycline eye ointment and/or oral tetracycline or azithromycin is effective and helps to prevent secondary cases. Topical therapy alone may not eradicate infection. Surgery to correct lid deformities may prevent blindness.

Prevention

- Mass treatment with tetracycline ointment or oral azithromycin
- Improving personal hygiene and general sanitation.

Streptococcal tonsillopharyngitis

Epidemiology

Sore throats are common in the general population and *S. pyogenes* is the cause in one-quarter of cases.

- Infection is usually sporadic, but periodic outbreaks occur
- The peak incidence is in school-age children during winter
- Transmission is by upper respiratory droplets and is facilitated by overcrowding. Rarely, milk-borne outbreaks occur
- *Streptococcus pyogenes* pharyngeal colonisation occurs in 15–20 per cent of children; in adults it is considerably lower
- Group C and G β-haemolytic streptococci occasionally cause tonsillitis
- The IP is 2–5 days.

Pathology and pathogenesis

Streptococcus pyogenes contains or produces a variety of intra- and extracellular toxins which enhance virulence. The secreted proteins include:

- pyrogenic exotoxins (A, B and C (responsible for scarlet fever and toxaemia))
- streptolysins O and S (responsible for β-haemolysis)
- deoxyribonucleases, hyaluronidase, and streptokinase which facilitate the spread of bacteria through tissues.

Antibodies to these products are used in the serodiagnosis of recent streptococcal infection. Of the intracellular and cell-wall products, the M-protein is the major virulence antigen. It is also the major antigen by which strains are typed, and over 80 such serotypes are recognised. This is epidemiologically important because certain serotypes are associated with streptococcal diseases:

- scarlet fever (types 1, 3 and 4)
- glomerulonephritis (types 12, 49, 55, 57 and 60)
- rheumatic fever (types 1, 3, 6 and 18).

Clinical features

Distinction of streptococcal tonsillitis from other causes is often difficult. Suppurative and post-infectious complications of streptococcal infection can be prevented or attenuated by antibiotics, whereas their inappropriate use in glandular fever may result in a severe skin rash. It is therefore important to make a clinical diagnosis. Typical of streptococcal tonsillopharyngitis are:

- an abrupt onset with sore throat, fever and headache
- vomiting and abdominal pain in children
- white exudate over swollen tonsils
- tender upper cervical lymphadenitis
- a self-limiting illness in treated, uncomplicated cases lasting 3–5 days
- the occurrence of a scarlitiniform rash in 5–10 per cent of cases.

Complications

Suppurative

- Peritonsillar (quinsy) and retropharyngeal abscesses
- Otitis media and sinusitis
- Suppurative cervical lymphadenitis
- Jugular venous and intracranial sinus thromboses.

Non-suppurative

- Scarlet fever
- Streptococcal toxic shock
- Rheumatic fever
- Glomerulonephritis
- Erythema nodosum
- Henoch–Schönlein purpura.

Diagnosis

Most sore throats are viral. Exudative tonsillitis may also be due to *C. diphtheriae* (now rare), EBV and adenovirus (when it may be associated with conjunctivitis – pharyngoconjunctival fever). Diagnosis is confirmed by:

- throat swab culture on blood agar (β-haemolysis and reaction with group A-specific antibody)
- a four-fold rise or fall or single high-titre antibody to one of the major antigenic constituents (usually ASO antibodies).

Treatment

Therapy is intended to prevent complications. First-line treatment is penicillin: group A streptococci remain universally susceptible. Where there is a history of penicillin allergy, erythromycin is a suitable alternative, although 5 per cent of strains are resistant. When the clinical features are severe or suppurative complications have intervened, IV benzylpenicillin should be administered.

Prevention

Recurrent tonsillitis

• Tonsillectomy is not of benefit except in a few children with recurrent, bacteriologically proven infections.

Rheumatic fever

• Initial treatment course of 10 days of oral penicillin
• Continuous prophylaxis with oral penicillin or monthly IM benzathine penicillin
• M-protein vaccine is under development.

Prognosis

Fatalities are exceptionally rare but may occur in toxin-mediated (scarlet fever and toxic shock syndrome) and severe suppurative (septicaemia and intracranial sinus thromboses) complications.

Diphtheria

Epidemiology

Diphtheria is a severe infection characterised by membrane formation in the throat and toxaemia which damages heart muscle and nerve tissue. *Corynebacterium diphtheriae* is characterised as follows:

• it is differentiated into three bacteriological types – *gravis*, *intermedius* and *mitis*; all are capable of severe disease
• it is transmitted by droplet from a nasopharyngeal case or carrier; occasionally skin infection occurs when direct contact is important
• it is now rare in the UK (>90 per cent immunised, with <5 cases/year)
• it rarely affects adults unless unimmunised; most cases are imported
• it has an IP of 2–5 days.

Pathology and pathogenesis

Corynebacterium diphtheriae is relatively non-invasive and causes a mild inflammatory reaction in the tonsil. Virulence results from the production of a potent exotoxin which inhibits protein synthesis by interference with mRNA. Locally this causes epithelial necrosis, an adherent membrane and surrounding oedema. As more toxin is produced it is absorbed from the membrane into the bloodstream where it affects the heart and nerves. Toxin is readily absorbed from the throat but only slightly from the nose, larynx or skin.

Clinical features

Clinical features depend upon the site of membrane formation. In tonsillopharyngeal disease:

- sore throat and low-grade fever commence gradually
- the membrane first appears on one or both tonsils and may spread to the pharynx, palate or buccal mucosa
- the membrane is off-white and thick; the extent of membrane formation correlates with severity
- there may be associated marked cervical adenitis and oedema, producing the classical 'bullneck'.

In laryngeal disease:

- hoarseness, a croupy cough and stridor develop
- with time, inspiratory recession of tissues and cyanosis will occur.

In anterior nasal disease:

- blood-stained unilateral nasal discharge occurs
- symptoms of toxicity are mild.

In cutaneous infection:

- chronic cutaneous ulcers develop, with a grey membrane
- symptoms of toxicity are mild
- the ulcers are a reservoir of *C. diphtheriae* that can lead to pharyngeal infection/carriage.

Complications

- Myocarditis
- Neuritis
- Glomerulonephritis.

Two-thirds of patients develop myocarditis but in only 10–25 per cent of these is this clinically important. It occurs between days 10 and 20; the ECG may reveal ST–T wave changes and heart block. Neuritis affects 75 per cent of patients with severe disease and results from demyelination. It afflicts cranial and peripheral nerves in a sequential order: palate (3 weeks), oculomotor (4 weeks), respiratory (7 weeks) and peripheral (10 weeks). All paralyses recover in time.

Diagnosis

The major differential diagnoses are glandular fever and streptococcal tonsillitis. Laryngeal diphtheria can be confused with epiglottitis, croup and foreign-body obstruction. Confirmation of diagnosis is by culture from throat and nasal swabs but treatment must not be delayed. Because non-toxigenic *C. diphtheriae* are not infrequently recovered from throat and nasal swabs, toxin studies must be performed on the isolate.

Treatment

The outcome is improved by prompt initiation of antitoxin therapy using horse hyperimmune antiserum. The dose depends on the site and severity of disease. Anaphylactic reactions may occur, especially in patients previously treated with this product. Erythromycin should be given to eradicate *C. diphtheriae* and post-treatment clearance swabs checked.

Prevention

Immunisation at 2, 3 and 4 months, and at entry into primary school. Close contacts of the index case should have:

- their throats examined by an experienced physician
- nose and throat swabs taken for *C. diphtheriae* culture
- a course of erythromycin prophylaxis
- their immunisation status assessed. If non-immune or of uncertain immunity, they should receive a primary course of diphtheria vaccine. If previously immunised, they should receive a reinforcing dose of adult diphtheria vaccine.

The Schick test is no longer available for determining immunity.

Prognosis

Case fatality rate is 5–10 per cent. The causes of death are severe toxaemia or laryngeal obstruction in the early days, heart failure in the 2nd or 3rd week, or respiratory failure at the 6th week.

Glandular fever

Epidemiology

Infection with EBV is almost universal. In childhood, primary EBV infection is usually nonspecific or asymptomatic, whereas in adolescents and adults glandular fever usually ensues. In EBV infection:

- 50 per cent of children in Western countries are infected by 5 years of age
- earlier acquisition occurs in the poor and developing nations
- the virus is excreted intermittently in most seropositive persons following primary infection
- transmission is by direct or indirect (via hands) saliva contact; rarely, infection follows blood transfusion.

In glandular fever:

- sporadic disease occurs with equal sex distribution
- the incidence is 50/100,000 in Western countries
- only 6 per cent have a history of contact with another case
- the incubation period is 2–6 weeks.

Pathology and pathogenesis

EBV is a herpesvirus. After oropharyngeal inoculation, EBV infects local epithelial cells and B-lymphocytes. Dissemination of infected B-lymphocytes then follows with subsequent production of a T-cell response characterised by atypical mononucleosis. This results in lymphoid hyperplasia with lymphadenopathy, splenomegaly and hepatomegaly. After infection, cells that contain the EBV genome are capable of continuous *in vitro* cultivation (immortalisation).

Clinical features

The classic triad of sore throat, fever and lymphadenopathy follows on from a prodrome of fever, tiredness, vague malaise and headache. The important clinical findings are:

- exudative tonsillitis with peritonsillar oedema
- petechiae at the junction of hard and soft palate
- cervical, axillary and inguinal lymphadenopathy with discrete non-tender glands
- splenomegaly (50 per cent), hepatomegaly (15 per cent) and jaundice (5 per cent)
- maculo-papular pruritic rash (90 per cent) after antibiotics (especially ampicillin/amoxycillin). Occasionally a rash may occur without preceding antibiotics (5 per cent).

Complications

The vast majority of patients with glandular fever recover uneventfully. Complications are rare but include:

- pharyngeal and tracheal obstruction
- Coombs' positive haemolytic anaemia, thrombocytopenia
- meningoencephalitis, Guillain–Barré syndrome
- splenic rupture, severe hepatitis, myocarditis and pneumonitis
- chronic fatigue.

True relapses of glandular fever are rare.

Diagnosis

Characteristic haematological findings are: leucocytosis, relative and absolute lymphocytosis, the presence of atypical lymphocytes, and a positive heterophile antibody test (Paul–Bunnell or Monospot (90 per cent of cases)). When considering the diagnosis:

- the heterophile antibody test may be negative early in infection, in children, in the elderly and in mild cases
- differential absorption with guinea-pig kidney cells and ox red cells distinguishes glandular fever from heterophile antibodies associated with serum sickness
- false-positive heterophile antibody tests occasionally occur in hepatitis A, malaria, rubella and lymphoma
- positive heterophile antibody tests persist for many months
- atypical lymphocytosis also occurs in CMV, HIV, human herpesvirus 6 (HHV-6), mumps, toxoplasma and hepatitis virus infections
- heterophile antibody negative glandular fever may result from EBV, HIV, CMV or *Toxoplasma* infections
- estimating IgM antibody to viral capsid antigen is useful in confirming the diagnosis in heterophile antibody negative cases
- mildly deranged transaminases occur in 90 per cent.

Treatment

A short course of steroids reduces tonsillopharyngeal oedema and is indicated when there is impending airway obstruction. They may also be useful when severe thrombocytopenia or acute haemolysis complicates infection.

Prevention

- Isolation in hospital is unnecessary
- No vaccine is available.

Prognosis

Practically all cases make a full recovery. Death in an immunocompetent person is rare but may result from respiratory obstruction, ruptured spleen or encephalitis. Fatalities may also occur in patients with inherited T-cell defects. EBV has been associated with several malignancies:

- nasopharyngeal carcinoma in China
- Burkitt's lymphoma in equatorial Africa
- immunoblastic lymphoma in X-linked lymphoproliferative syndrome
- B-cell lymphomas in the immunocompromised
- cerebral lymphoma, oral hairy leucoplakia and lymphocytic pneumonitis in HIV-infected patients.

Mumps

Epidemiology

Mumps is a generally mild and self-limiting acute infection caused by a member of the paramyxovirus group. Humans are the only natural hosts and infection is common and widespread.

- 85 per cent of adults show serological evidence of past infection
- Schoolchildren and adolescents are principally affected, although more cases are now being seen in adults
- Clinical mumps is rare in children <2 years of age
- Cyclical (2–5 years) and seasonal (spring) peaks occur
- Transmission is by droplet spread or direct/indirect contact with saliva
- Over 95 per cent reduction in incidence has occurred where attenuated mumps vaccine has been in use for 10 years
- Infectivity lasts from a week before salivary gland swelling starts to up to 9 days after, with a peak just before and at the onset of the parotitis
- The IP is 17–19 days, with limits of 12–25 days.

Pathology and pathogenesis

Initially the virus replicates in the epithelium of the upper respiratory tract, followed by viraemic dissemination to other organs including salivary glands and meninges.

Clinical features

The illness presents with a nonspecific prodrome of headache, sore throat, malaise and fever lasting 1–2 days. Salivary gland inflammation and enlargement then follows. Parotitis:

- is present in 70 per cent of cases
- starts unilaterally but involves the other gland after 1–5 days in 75 per cent of cases
- is heralded by earache and discomfort/tenderness at the angle of the jaw, which is aggravated by chewing
- causes rapid, painful and tender enlargement, obscuring the angle of the jaw
- is associated with involvement of the other salivary glands in 10 per cent
- resolves over 1–2 weeks
- may result in difficulty in opening the mouth (trismus)
- is associated with swelling and redness at the opening of the parotid duct opposite the upper second molar.

Complications

Neurological

- Meningitis
- Encephalitis
- Nerve deafness, facial palsy, Guillain–Barré syndrome, transverse myelitis.

Mumps meningitis is second to enterovirus as a common cause of viral meningitis and may be the presenting syndrome when only half have clinical evidence of parotitis. It occurs in 10 per cent of those presenting with parotitis and can precede salivary gland involvement, or follow it by as long as 2 weeks; usually it starts 4 days after parotitis. It is associated with a lymphocytic cerebrospinal fluid (CSF) with raised protein, normal or low (5–10 per cent) glucose and mumps virus isolation. Encephalitis is rarer, usually presents in the later stage of infection, and has a significant morbidity and mortality.

Glandular

- Epididymo-orchitis, oophoritis
- Pancreatitis.

Epididymo-orchitis occurs in 25 per cent of postpubertal males and is unilateral in 85 per cent and appears as other signs settle. It is heralded by the abrupt return of fever and malaise and indicated by rapid painful swelling of the testes and reddening of the overlying scrotal skin. Occasionally it is associated with atrophy of the testis; sterility is very rare. Rarely, it is the only manifestation of mumps. Oophoritis occurs much less frequently.

Other complications

- Arthritis, myocarditis, pericarditis.

Diagnosis

In typical cases, the diagnosis is clinical and simple, but in atypical cases or where there are associated complications, laboratory confirmation is necessary. Circumstantial evidence of recent mumps infection is supported by:

- a normal or low WCCt with a relative lymphocytosis
- an elevated serum amylase
- a lymphocytic CSF with raised protein and normal or low sugar.

Proof of infection is provided by:

- isolation of the virus from CSF, saliva or urine
- demonstration of a four-fold rise or fall, or single high-titre antibody to mumps 'S' (0–12 weeks) or 'V' (2 weeks – years) antigens.

Treatment

- Symptomatic — mild analgesics, mouth washes and easily chewed diet
- In orchitis, strong analgesics, local ice packs and scrotal support are comforting.

Prevention

- In hospital, patients should be isolated
- Immunisation of susceptible contacts following exposure is not protective. Since subclinical infection frequently occurs, a non-immune contact should be considered infectious from 12–25 days after exposure
- Since 1988, attenuated live mumps vaccine as a component of combined MMR vaccine has been in routine use in the UK. The vaccine is recommended for all children aged 12–15 months unless contraindicated, and should also be offered to older, unimmunised children before they start school. The immunity appears to be long lasting and the vaccine is well tolerated. Transient swelling of the parotid glands lasting for less than 24 hours may develop in the 3rd week.

Prognosis

Apart from a very rare death from encephalomyelitis, there is no mortality from mumps. Permanent sterility or deafness are very rare sequelae.

Epiglottitis

Epidemiology

Acute epiglottitis is a rapidly progressive life-threatening infection resulting from capsulate *H. influenzae* type b. It is a medical emergency which:

- affects children of 2–5 years of age
- is transmitted by aerosol from a carrier or a case
- has an IP of 1–4 days
- has become rare with the introduction of *H. influenzae* type b (Hib) vaccine.

Pathology and pathogenesis

Capsulate strains of *H. influenzae* are invasive by virtue of surface pili (aid adherence and mucosal penetration), lipopolysaccharide (inhibits ciliary beating) and the capsule (assists colonisation and resistance to neutrophils). Infection is infrequent but invariably results in severe disease (meningitis, epiglottitis, septicaemia, cellulitis, pneumonia). By contrast, non-capsulate strains commonly cause local complications (otitis media, sinusitis, conjunctivitis, bronchopneumonia) but are very rarely invasive. Following colonisation and local invasion, a brisk inflammatory reaction develops in the epiglottis, with rapid swelling resulting in airways obstruction.

Clinical features

Epiglottitis progresses rapidly and may be fulminant. Typically, children:

- have no prodromal illness
- present sitting, leaning forward and drooling secretions
- are unable to swallow or talk
- have a weak cough and inspiratory stridor
- are tachycardic, hypotensive, febrile and toxic-looking
- rapidly progress to airways obstruction.

The throat must *not* be examined until an anaesthetist is present and facilities are available to secure the airway. Attempts to do so may result in complete airway obstruction or vagally mediated cardiopulmonary arrest.

Complications

These may be local (airways obstruction, hypoxaemia) or systemic (septicaemic shock). Because complications are the rule and not the exception, the child must be admitted direct to the intensive care unit.

Diagnosis

The diagnosis is clinical. Typically, there is:

- a cherry-red, swollen epiglottis visible at intubation
- leucocytosis with neutrophilia
- *Haemophilus influenzae* type b grown from blood and epiglottal cultures
- a swollen epiglottis on lateral neck X-ray.

Treatment

- Establish the airway
- Give IV cefotaxime. Nearly one-third of *H. influenzae* type b isolates are resistant to ampicillin and 1–3 per cent to chloramphenicol.

Prevention

- Hib vaccine at 2, 3 and 4 months
- The index case should receive rifampicin 20 mg/kg/day daily for 4 days before hospital discharge
- Close family contacts where there is a child <5 years should also receive rifampicin.

Prognosis

With early recognition, full recovery is usual. However, 5–10 per cent of cases are fulminant and are often dead on arrival at hospital.

Laryngotracheobronchitis

Epidemiology

Acute laryngotracheobronchitis (croup) is an acute infection of the respiratory tract, particularly involving the subglottic area, resulting in the characteristic stridor. Croup:

- is most commonly caused by parainfluenzae type I
- occurs in winter epidemics in 2-yearly cycles
- affects children between 3 months and 3 years of age
- can cause second infections
- affects boys more commonly
- is self-limiting; complications are rare
- has an IP of 2–4 days.

Pathology and pathogenesis

The cardinal features of croup (stridor, hoarseness and cough) result from viral-induced inflammation and oedema of the larynx and trachea, although the whole respiratory tract is involved in the infection. Hypoxaemia results from parenchymal lung inflammation and laryngeal obstruction.

Clinical features

Characteristics are as follows:

- there is a short prodrome of coryza, sore throat and cough
- hoarseness, barking cough and stridor follow
- airways obstruction develops with recession of the soft tissues of the neck and abdomen on inspiration and, in severe cases, cyanosis
- fever is not striking and the WCCt is normal or low
- after a fluctuating course lasting 4–5 days, the symptoms settle
- airways obstruction is very rare in older children and adults.

Complications

- Pneumonia
- Respiratory failure, pneumothorax, pulmonary oedema and aspiration pneumonia
- Airway hyperreactivity.

Diagnosis

The diagnosis is clinical. Viral aetiology is confirmed by:

- viral antigen detection on nasopharyngeal aspirates
- culture.

Treatment

- Oxygen if hypoxaemic (monitor gases or pulse oximetry)
- Humidification
- There is no specific antiviral therapy
- Corticosteroids should be used in severe cases.

Prevention

No vaccine exists.

Prognosis

The illness is benign. Fatalities result from respiratory failure or one of the other major acute complications.

Common cold

Epidemiology

The common cold is a mild self-limiting catarrhal illness associated with a low-grade fever. It is not a single entity but a clinical syndrome with many viral causes. It is a major cause of lost school and work days.

- Rhinoviruses account for one-third of cases; coronaviruses, adenoviruses, parainfluenza and influenza viruses, RSV and enteroviruses for another third; in the remaining it is impossible to identify the cause
- Secondary bacterial infection resulting in otitis media, sinusitis or tracheobronchitis occurs in 2–3 per cent
- Reinfections are common and, on average, an adult suffers two or three, and a child six to eight colds per year
- Transmission is by droplet infection or direct contact, usually via hands
- Annual epidemics occur during the cold or wet seasons
- The IP is 12 hours to 3 days.

Pathology and pathogenesis

Limited damage occurs following viral invasion of the columnar epithelial cells with rhinoviruses. CT scans have demonstrated thickened nasal walls, engorged turbinates and fluid or mucosal thickening in the sinuses indicating that the common cold is not a localised infection. Pharyngitis and conjunctivitis may occur with adenoviral or enteroviral infection.

Clinical features and complications

The illness starts with slight fever, malaise and irritation of the nasal mucosa and pharynx, soon followed by a profuse watery nasal discharge, repeated sneezing and coughing. Later the nasal discharge becomes more purulent and the mild systemic symptoms subside. The only complication is secondary bacterial infection (*H. influenzae*, *S. pneumoniae*, *S. pyogenes*) leading to sinusitis, otitis media or tracheobronchitis.

Diagnosis, treatment and prevention

Rarely is there any indication to confirm the aetiology of a common cold, which is a clinical diagnosis. Nose and throat swabs for viral culture, direct antigen testing on nasopharyngeal aspirates and serology are available. Treatment is symptomatic with aspirin gargles for a sore throat, decongestants for a blocked nose and antibiotics only if secondary bacterial infection has occurred. Interferon A has a modest benefit on symptoms; there is no vaccine.

Tuberculous lymphadenitis

Epidemiology, pathology and clinical features

- Cervical and mediastinal glands are affected most commonly, followed by axillary and inguinal: in 5% more than one regional group is involved
- Signficant constitutional disturbance and evidence of associated tuberculosis is usually lacking
- In non-immigrant children in the UK, the majority of mycobacterial cervical lymphadenitis is caused by non-tuberculous mycobacteria (e.g. *Mycobacterium avium*, *Mycobacterium scrofulaceum*)
- In immigrants to the UK, lymphadenitis (particularly cervical) is the commonest non-pulmonary presentation
- The lymph nodes are painless, initially mobile, and occasionally discharge through the skin causing a 'collar-stud' abscess.

Diagnosis

- Ziehl–Nielsen microscopy (positive in only 25 per cent)
- Histology revealing caseating granulomata
- Culture on Lowenstein–Jensen or Bactec media.

Treatment and prevention

- Short-course chemotherapy (as for pulmonary tuberculosis (see p. 99))
- Paradoxical enlargement as a result of a hypersensitivity reaction occurs occasionally during or even after completing therapy
- Surgical excision is sometimes necessary
- In some cases of mediastinal lymphadenitis where the clinical picture is very suggestive, empirical therapy is often given.
- Prevention as for pulmonary tuberculosis (see p. 99).

CHAPTER 7

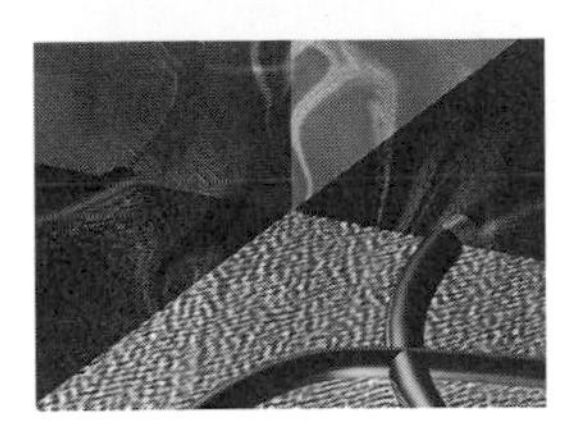

Lower Respiratory Tract Infections

CLINICAL SYNDROMES

Pleural effusion, empyema and lung abscess

Major presenting features: *chest pain, cough, fever.*

Causes

Pleural effusion and empyema

Post-pneumonic: STREPTOCOCCUS PNEUMONIAE, *Staphylococcus aureus, Streptococcus pyogenes, Mycoplasma pneumoniae, Legionella pneumophila*

Primary: STREPTOCOCCUS MILLERI, ANAEROBES, MICROAEROPHILIC STREPTOCOCCI, *Escherichia coli* and other 'coliforms' (often polymicrobial); *Mycobacterium tuberculosis, Actinomyces, Nocardia.*

Lung abscess

Necrotising pneumonia: S. AUREUS, *Klebsiella pneumoniae, Fusobacterium necrophorum, S. pyogenes, Pseudomonas aeruginosa, L. pneumophila*

Aspiration: ANAEROBES, MICROAEROPHILIC STREPTOCOCCI, S. MILLERI (often polymicrobial)

Others: M. TUBERCULOSIS, *Entamoeba histolytica*, actinomycosis, *Nocardia, Pseudomonas pseudomallei*, aspergilloma.

Distinguishing features

	Pleural effusion	*Empyema*	*Lung abscess*
Fever	normal/low	high/hectic	high/hectic
Chest pain	present	present	absent
Clubbing	absent	if chronic	if chronic
CXR			
site	costophrenic	costophrenic or loculated	lungs
air/fluid level	absent	rare	usually present
Full blood count			
WCCt	normal	raised, neutrophilic	raised, neutrophilic
anaemia	absent	may be present	may be present
Pleural/abscess fluid			
WCCt increase	moderate, lymphocytic	high, neutrophilic	high, neutrophilic
protein	moderate rise	very high	very high
Gram stain	rarely positive	usually positive	usually positive
culture	negative	positive	positive

Complications

Lung abscess: bronchopleural fistula, empyema, brain abscess, bronchiectasis

Empyema: empyema necessitans, septicaemia, pericarditis.

Investigations

- Full blood count (FBCt), differential white cell count (WCCt), erythrocyte sedimentation rate (ESR)
- Chest X-ray (CXR) (Fig. 7.1): further imaging may be necessary (e.g. ultrasound scan/computerised tomography (CT) scan)
- Blood culture (occasionally positive in empyema/lung abscess)
- Fluid from pleural cavity/lung abscess:
  - microscopy
  - Gram and Ziehl–Nielsen (ZN) stains
  - culture: aerobic and anaerobic; *M. tuberculosis*
  - protein concentration
  - antigen testing for *S. pneumoniae* (if negative culture).

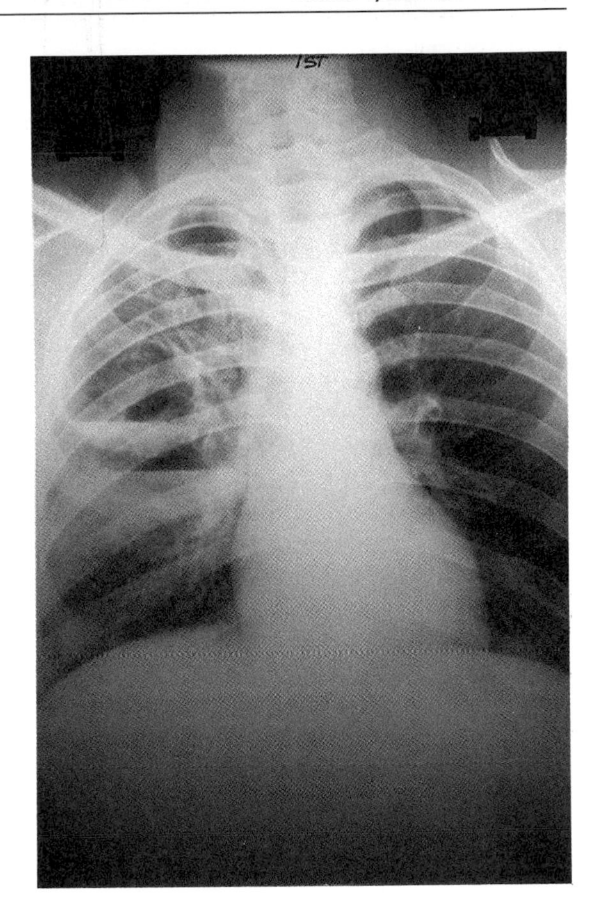

Fig. 7.1. Anaerobic lung abscesses.

Differential diagnoses

Pleural effusion/empyema

- Neoplasm, pulmonary infarct, congestive cardiac failure, autoimmune disease (systemic lupus erythematosus (SLE) rheumatoid disease, etc.), hypoalbuminaemia, subdiaphragmatic infection, oesophageal perforation, pancreatitis.

Lung abscess

- Neoplasm, pulmonary infarct, infection distal to bronchial obstruction, septic embolus.

Treatment

Empyema and lung abscess

- Aspirate/drain fluid where feasible

- Initial parenteral antibiotics, guided by isolate(s) and sensitivity patterns
- Post-pneumonic: treat underlying condition:
  - *S. aureus:* flucloxacillin
  - *K. pneumoniae:* cefotaxime
  - *S. pyogenes:* benzylpenicillin
  - *M. pneumoniae:* erythromycin
  - *S. pneumoniae:* cefotaxime
- Primary:
  - anaerobes: metronidazole
  - *S. milleri:* benzylpenicillin
  - coliforms: cefotaxime
  - consider co-amoxiclav or clindamycin, as often polymicrobial
- *Mycobacterium tuberculosis*: isoniazid, rifampicin and pyrazinamide.

Adult community-acquired pneumonia

Major presenting feature: *cough, fever, chest pain.*

Causes

- STREPTOCOCCUS PNEUMONIAE, S. AUREUS, HAEMOPHILUS INFLUENZAE, *S. pyogenes, K. pneumoniae, Moraxella catarrhalis*
- MYCOPLASMA PNEUMONIAE, L. PNEUMOPHILA, COXIELLA BURNETII, CHLAMYDIA PSITTACI, CHLAMYDIA PNEUMONIAE ('atypical' pneumonia agents)
- *Neisseria meningitidis, Yersinia pestis, Chlamydia trachomatis*, other *Legionella* spp.

Distinguishing features

	Pneumococcal pneumonia	*'Atypical' pneumonia*
History	sometimes a preceding viral illness	contact with birds, animals, affected family or water. May be seasonal or outbreak
IP	1–3 days	usually >7 days
Onset	sudden with rigor	may be gradual
Influenza-like prodrome	brief or absent	present
Pleuritic chest pain	usually present	often absent
Cough	starts early, <2 days	starts late, >5 days
Sputum	initially dry, then rusty coloured	initially dry, then mucopurulent

	Pneumococcal pneumonia	*'Atypical' pneumonia*
Haemoptysis	very rare	occasional
Non-respiratory symptoms	uncommon	common
Splenomegaly	rare	not uncommon
Rash	rare	15 per cent with *M. pneumoniae*
Total WCCt ($\times 10^9$/litre)	usually >15	usually $\leqslant 15$
Hyponatraemia	occasional	marker for *L. pneumophila*
Hepatitis	rare	occasional
CXR	well defined, lobar or lobular, unilateral	patchy, poorly defined, bilateral
Sputum Gram stain	Gram-positive cocci	no organisms
Response to β-lactam antibiotics	excellent, rapid	nil
Response to erythromycin	excellent, rapid	excellent, slow
Main diagnostic method	microscopy, culture or antigen detection	serology

Complications

Respiratory failure is a rare but occasional complication of all the respiratory pathogens.

Streptococcus pneumoniae: pleural effusion, empyema, bacteraemia, pericarditis, endocarditis, meningitis

Mycoplasma pneumoniae: pleural effusion, bullous myringitis, haemolysis, encephalitis

Legionella pneumophila: pleural effusion, empyema, lung fibrosis, neuropathies

Coxiella burnetii: hepatitis

Chlamydia psittaci: hepatitis, encephalopathy, myocarditis.

Investigations

- FBCt with differential WCCt
- Biochemical profile (liver function tests (LFTs), albumin and urea/ creatinine)
- CXR
- Arterial gases or pulse oximetry
- Sputum:
  - Gram stain

- antigen or DNA detection
- culture (including *Legionella*)
- Blood culture (*S. pneumoniae* septicaemia occurs in 25 per cent)
- Antigen detection:
  - *S. pneumoniae* (serum, sputum and urine)
  - *L. pneumophila* (sputum, urine)
- Acute and convalescent serum: all atypical pathogens.

Differential diagnoses

- Rarely, causes of chronic pneumonia can present acutely (see below)
- Pulmonary infarction, pulmonary oedema, allergic or cryptogenic alveolitis
- Collapse associated with foreign body, lung cancer.

Treatment

Streptococcus pneumoniae

- Intravenous (IV) benzylpenicillin (penicillin-sensitive)
- IV cefotaxime (penicillin intermediate-resistant)
- IV cefotaxime and IV vancomycin (penicillin-resistant).

'Atypical' pneumonia

- Oral or IV erythromycin
- IV rifampicin or IV ciprofloxacin in addition for severe Legionnaires' disease
- Oral or IV tetracycline is an alternative if not *Legionella*.

Severe community-acquired pneumonia, aetiology unknown

- IV erythromycin and IV cefotaxime (assuming high-level penicillin resistance rare in community isolates).

Prevention

Streptococcus pneumoniae

- Vaccination for high-risk groups
- Antibiotic prophylaxis (children with sickle cell anaemia).

Chronic pneumonia

Major presenting features: *chronic cough, fever, night sweats, weight loss.*

Causes

- MYCOBACTERIUM TUBERCULOSIS, *M. bovis, M. africanum*
- Mixed anaerobic/aerobic bacteria

- Atypical mycobacteria, *Nocardia*, *P. pseudomallei*, *Actinomyces*
- *Cryptococcus neoformans*, *Aspergillus*, histoplasmosis, coccidioidomycosis
- *Paragonimus westermani.*

Distinguishing features

	Tuberculosis	*Acute community-acquired pneumonia*
Epidemiology	old; immigrant	any age, any ethnic group
History	recent or remote contact with tuberculosis	preceding viral illness or contact with birds, animals, water or affected family
Onset	subacute/chronic	acute
Fever	low grade	high and hectic
Weight loss	moderate	minimal
Haemoptysis	not uncommon, rarely massive	occasional
WCCt	normal	raised in *S. pneumoniae*
Haemoglobin	anaemia	normal
CXR		
cavitation	usual in adults	uncommon (seen in *S. aureus* and *K. pneumoniae*)
apical lesion	usual	uncommon
distribution	commonly bilateral, patchy	usually unilateral and lobar
miliary	occasional	never
Response to therapy for acute pneumonia	nil	good

Complications

Mycobacterium tuberculosis: respiratory failure, massive haemoptysis, lung fibrosis, pleural effusion/empyema, meningitis, focal extrapulmonary disease

Acute community-acquired pneumonia: see p. 68.

Investigations

Chronic pneumonia

- FBCt, differential WCCt, ESR
- Biochemical profile (LFTs, albumin, urea/creatinine)
- Pulse oximetry: arterial gases may be indicated
- CXR
- Sputum microscopy:
  - ZN and/or auramine
  - Gram
  - fungal
- Sputum culture:
  - Lowenstein–Jensen, Bactec (for tuberculosis)
  - fungal
  - anaerobic (*Actinomyces*)
  - prolonged standard (*Nocardia* and *Actinomyces*)
- If ZN is negative on three samples but appearances are consistent with tuberculosis, consider:
  - induced sputum
  - bronchoscopy
  - gastric washings
  - polymerase chain reaction (PCR) on specimens
- Serum: cryptococcal antigen.

Differential diagnoses

- Neoplasia, sarcoidosis, Wegener's granulomatosis, pulmonary embolus.

Treatment

Tuberculosis

- Isoniazid (6 months), rifampicin (6 months), pyrazinamide (2 months); ethambutol (2 months) should be added initially if resistance likely (past treatment, recent immigrant).

Prevention

- Bacille Calmette–Guérin (BCG) vaccine
- Notification, screening of close contacts, and primary prophylaxis of persons with primary subclinical infection.

Bronchiolitis and childhood pneumonia

Major presenting features: *cough, breathlessness, fever.*

Causes

Bronchiolitis

- RESPIRATORY SYNCYTIAL VIRUS (RSV)
- Parainfluenza, influenza virus, adenovirus, rhinovirus, *M. pneumoniae.*

Childhood pneumonia

- RSV, measles, varicella zoster, parainfluenza, influenza, adenovirus, coxsackie
- *STREPTOCOCCUS PNEUMONIAE, H. INFLUENZAE* (TYPE B), *S. aureus, M. pneumoniae*
- *Mycobacterium tuberculosis.*

Distinguishing features

	Bronchiolitis	*Bacterial pneumonia in childhood*
Epidemiology		
season	winter, early spring epidemics	commoner in winter, can occur anytime
age	infants	usually <5 years
Prodromal upper respiratory tract symptoms	present	often absent
Fever	low grade	high
Toxicity	usually mild	marked
Wheeze	present	usually absent
Recession	present	absent
Auscultation	diffuse wheeze with crackles	localised crackles or consolidation
WCCt	normal or slightly raised	raised
WCCt differential	normal/lymphocytic	neutrophilic
CXR		
hyperinflation	present	absent
abscesses or pleural effusion	absent	may be present
Blood culture	negative	may be positive

Complications

Bronchiolitis: respiratory failure, apnoeic attacks, secondary bacterial pneumonia, recurrent wheezing, bronchiolitis obliterans

Bacterial pneumonia: abscesses, cavities, pneumatocoeles (*S. aureus*), pleural effusion, empyema, bacteraemia, metastatic abscess, meningitis.

Diagnosis

- FBCt, differential WCCt
- Biochemical profile (LFTs, albumin, urea/creatinine)
- CXR
- Pulse oximetry or arterial gases
- Blood culture
- Throat swabs for viral and bacterial culture
- Nasal swab for viral culture
- Nasopharyngeal aspirate for RSV and other viral antigens
- Serum:
  - acute and convalescent sera (atypical agents)
  - pneumococcal antigen detection
- Urine: pneumococcal antigen detection.

Differential diagnoses

- Asthma, heart failure, congenital heart disease (L–R shunt), metabolic acidosis.

Treatment

Bronchiolitis

- Humidifed oxygen, oral nutrition
- Nebulised ribavirin in those likely to develop severe disease: (infants <2 months, those born prematurely or with chronic cardiorespiratory disease, and immunocompromised children).

Bacterial pneumonia

- Humidifed oxygen
- IV cefotaxime alone, or with IV erythromycin (if atypical agent possible).

Prevention

- *Haemophilus influenzae* type b (Hib) vaccine
- Antibiotic prophylaxis with rifampicin for all close family contacts of index case of *H. influenzae* type b disease where there are other children aged <5 years in the family.

SPECIFIC INFECTIONS

Influenza

Epidemiology

Influenza is a common, highly infectious epidemic disease. It belongs to the myxovirus group and exists in two main forms, A and B. Influenza A is responsible for periodic worldwide epidemics (every 1–3 years) and unpredictable pandemics (every 1–2 decades).

- Four pandemics of influenza A have been recorded this century (1918, 1957, 1968 and 1977): the first of these killed 200,000 persons in England and Wales, and 20 million worldwide
- There is a higher incidence of influenza in the winter months with epidemics peaking in December/January
- Infectivity is very high and extends from shortly before symptoms start until shortly after the pyrexia settles
- The common occurrence of subclinical infections, and the short IP, contribute to the rapid spread of disease
- The mode of transmission is by droplet infection and by hands and recently contaminated articles (e.g. handkerchiefs)
- The highest mortality is amongst the elderly
- Outbreaks in the Southern hemisphere from May to September usually predict the type of virus appearing in the Northern hemisphere the following winter
- Overall attack rates are 10–20 per cent during an epidemic
- Influenza B causes sporadic infections and local epidemics about every 2nd year. The disease is usually mild and affects children particularly
- The IP is 1–3 days (range 12 hours to 5 days).

Pathology and pathogenesis

Influenza A has the capacity to develop new antigenic variants at irregular intervals.

- The virus contains two surface antigens, namely a haemagglutinin (H antigen) and a neuraminidase (N antigen)
- These may change, and this is the basis of classification, together with the site and year of isolation (e.g. A/USSR/77 H1N1)
- Immunity develops specifically to H and N antigens; a change in either antigen will result in a loss of previous immunity
- Major change (antigenic shift) gives rise to pandemics and results from

genetic reassortment between humans and the animal reservoir and the appearance of a novel H or N antigen (or both)

- Minor change (antigenic drift) gives rise to epidemics and results from the accumulation of random point mutations in the RNA and subsequent slight alteration of the H and/or N antigen
- Antigenic variation occurs less frequently with influenza B
- The virus is cytopathic to the respiratory tract.

In pneumonitis, cytopathic involvement of the ciliated columnar epithelium of the trachea and bronchi, with sloughing of the epithelium, is found progressing to diffuse haemorrhage and hyaline membrane formation in fatal cases; inflammatory infiltrate is minimal.

Clinical features

The illness starts abruptly with fever, rigors, headache, aching eyes, myalgia, sore throat and dry cough. Most symptoms settle after 2–5 days but the cough and malaise may persist for 1–2 weeks. Physical examination shows only pyrexia and reddening of the fauces. In epidemics, mild attacks with short-lived fever and less prominent symptoms are very common.

Complications

Respiratory

- Pneumonitis
- Laryngotracheobronchitis and bronchiolitis
- Otitis media, secondary bacterial pneumonia
- Exacerbation of chronic respiratory disease.

Influenza pneumonitis occurs in 5 per cent of cases. It is more likely in the elderly, in pregnant women and patients with underlying cardiac or pulmonary disease. Development of marked dyspnoea, dry cough and cyanosis is associated with hypoxia and diffuse fine interstitial shadows on CXR. In young children, influenza A or B may cause laryngotracheobronchitis or bronchiolitis. Secondary bacterial pneumonia (*S. aureus*, *S. pneumoniae* or *H. influenzae*) is an important cause of death. It is usually seen in the elderly and those with underlying respiratory disease. CXR reveals multiple areas of consolidation.

Non-respiratory

- Guillain–Barré syndrome, myelitis, encephalitis
- Reye's syndrome
- Myositis and myoglobinuria
- Myocarditis and pericarditis.

Diagnosis

In the presence of a confirmed outbreak, a patient with influenza-like symptoms is likely to have influenza. Confirmation can be achieved by demonstrating the following:

- virus from nose or throat swabs
- a four-fold rise or fall in antibody titre by complement fixation or haemagglutination assay
- antigen detection on pharyngeal aspirates
- specific nucleic acid (NA) amplification by PCR.

If secondary infection is suspected:

- blood culture and sputum Gram stain and culture should be performed.

Treatment

The severity of influenza is variable, but most patients require only symptomatic therapy.

- Amantadine will shorten the duration of symptoms by one-third if started within 48 hours. Treatment should be considered for symptomatic patients at high risk for complications
- Antibiotics (e.g. cefotaxime) with activity against the likely secondary pathogens (*S. aureus*, *S. pneumoniae* and *H. influenzae*) should be used in all ill patients with respiratory complications
- Aspirin should not be used in children because of the risk of Reye's syndrome.

Prevention

- Global monitoring of influenza activity allows forewarning of impending epidemics. Early detection of influenza-like illness in the UK is done through a network of 'spotter' general practitioners
- Influenza vaccine is recommended for persons with chronic respiratory, cardiac or renal disease, diabetes or other endocrine disorders, those with immunosuppression, old institutionalised persons, and for health-care workers during an epidemic
- Amantadine protects from symptomatic illness in 70 per cent of cases and should be considered for unvaccinated exposed persons at high risk for complicated disease.

Prognosis

The death rate increases with age, underlying illness and the presence of complications, particularly pneumonitis. The 1989/90 epidemic was thought to be responsible for 29,000 excess deaths in England and Wales.

Bronchiolitis

Epidemiology

Bronchiolitis is an acute lower respiratory tract infection (RTI) of infancy characterised by wheezing and hyperexpansion and resulting from inflamed, obstructed small airways. RSV is a major cause (60 per cent of cases) as are the parainfluenza viruses (20 per cent). Occasionally, influenza virus, adenovirus, rhinovirus, *M. pneumoniae* and enterovirus are responsible. RSV also causes tracheobronchitis, pneumonia and upper RTI with otitis media; asymptomatic infection is uncommon.

• RSV is an RNA paramyxovirus with two major subtypes, A and B

• Immunity is incomplete and repeated infections are common; severe disease rarely occurs after primary infection. By 2 years of age nearly all children have serological evidence of infection

• Epidemics occur annually in winter and early spring

• Disease is commoner in boys, breastfed babies and those from lower socioeconomic groups

• The most severe disease occurs in the youngest infants (<2 months old), those born prematurely or with chronic cardiorespiratory disease, and the immunocompromised

• Spread is by droplet infection to mucous membranes, either airborne or by direct contact; viral shedding lasts 1 week

• The IP is 3–5 days (range 2–8 days).

Pathology and pathogenesis

RSV causes patchy peribronchial inflammation with oedema, necrotic material and fibrin production leading to obstruction. If partial, this leads to air trapping and hyperinflation: if complete, to collapse. Interstitial pneumonia may also be present. Local immune response is probably important in pathogenesis as shown by raised specific immunoglobulin E (IgE) antibodies in nasopharyngeal secretions. However, systemic antibody is poorly protective because disease may occur despite maternal or vaccine-stimulated antibody.

Clinical features

The characteristics of RSV bronchiolitis are as follows:

• fever, coryza and cough precede the development of respiratory distress by 1–2 days

• pallor, tachypnoea, tachycardia and restlessness supervene

• supraclavicular, intercostal and subcostal recession of the soft tissues on inspiration develop, associated with expiratory wheeze

- auscultation reveals wheeze with or without diffuse crackles
- cyanosis, aggravated by coughing or by attempts at feeding, develops in severe cases
- fever is intermittent and rarely exceeds 39°C
- the disease lasts 3–7 days with gradual recovery over 1–2 weeks.

Complications

The major complication is respiratory failure, which is rare in previously healthy children. Up to two-thirds of fatal cases occur in patients with cardiopulmonary disease or who are immunosuppressed. Apnoea and hypoxia are not uncommon in hospitalised infants, whereas secondary bacterial infection is unusual. Hyperreactive airways and asthma may be linked with bronchiolitis in infancy. Bronchiolitis obliterans is a very rare and severe complication which appears to follow adenovirus bronchiolitis.

Diagnosis

The diagnosis is primarily epidemiological and clinical although the condition must be distinguished from asthma, which is usually recurrent, and from pneumonia, where the striking signs of airway obstruction are not normally seen. Nonspecific features supporting RSV include:

- normal or slightly raised WCCt with lymphocytic differential
- hyperinflation, prominent bronchial wall markings and multiple areas of collapse-consolidation on CXR
- hypoxaemia without hypercapnia (unless respiratory failure is developing).

Specific diagnosis is made through one of the following:

- virus isolation from a nose or throat swab
- antigen detection from nasopharyngeal aspirates.

Treatment

Children with signs of lower respiratory tract involvement are best treated in hospital in isolation. Severe cases should be:

- nursed in oxygen with monitoring of P_{O_2} or oxygen saturation
- nursed in a humid atmosphere
- tube fed if the infant is distressed by normal feeding
- sucked out repeatedly, which helps to maintain the airway
- carefully observed for overhydration which must be avoided
- considered for ventilation and aerosolised ribavirin if the disease is particularly severe.

Antibiotics, bronchodilators and corticosteroids are ineffective.

Prevention

- Control of cross-infection in children's wards
- Infants should not be unnecessarily exposed to adults and older siblings who are suffering from upper RTI
- Trials with killed RSV vaccine were discontinued in view of the occurrence of severe clinical disease in the vaccinated.

Prognosis

Deaths are exceptionally rare unless the infant has underlying disease, was born prematurely or is <2 months old. In these categories there is an appreciable mortality (5–35 per cent).

Chlamydia pneumoniae infection

Epidemiology

Chlamydia pneumoniae is the most recently identified and probably the commonest chlamydial pathogen of humans.

- It is a primary human pathogen without an animal reservoir
- Transmission is by aerosol
- Men are affected more frequently than women
- Antibodies are rare in children; by middle age, 50 per cent have antibodies
- Most infections are mild, subclinical infections are common and reinfections occur
- Pneumonia is the commonest manifestation of clinical disease, but bronchitis, 'influenza-like' illness, pharyngitis, sinusitis and otitis media also occur
- It is commonly associated with outbreaks and may be nosocomial
- It causes approximately 10 per cent of community-acquired pneumonias
- The IP is 10 days.

Pathology and pathogenesis

The organisms are small, coccoid bacteria which are obligate intracellular parasites. The pathogenesis is presumed to be very similar to that for *C. psittaci* (see p. 81).

Clinical features

Clinical distinction of pneumonia caused by *C. pneumoniae* from other aetiologies of pneumonia is difficult. Nevertheless, the following features are characteristic:

- pharyngitis and hoarseness
- a biphasic illness with initial improvement of upper respiratory tract symptoms
- the development of pneumonia 1–3 weeks later, with acute onset of dry cough and fever without chest pain
- a mild illness
- a relapsing course with persistent cough.

Complications

Severe disease leading to respiratory failure may occur in the elderly or those with underlying chest or heart disease. Chronic sinusitis sometimes occurs.

Diagnosis

The CXR usually has a single infiltrate, the WCCt is normal and the ESR is raised. Laboratory confirmation is by type-specific immunofluorescence assay. The standard chlamydial antibody test is the complement fixation text. This cannot differentiate *C. pneumoniae* from *C. psittaci* and lacks sensitivity, especially for reinfections. Antibody rises can be slow and blunted by prompt antibiotic therapy.

Treatment

Tetracycline and erythromycin are both effective against *C. pneumoniae*. The newer quinolones (e.g. ciprofloxacin) and macrolides (e.g. clarithromycin) may be effective alternatives. Treatment should be continued for 2 weeks.

Prevention

No effective vaccine is available.

Prognosis

The illness is benign. The very rare fatality results from respiratory failure.

Psittacosis

Epidemiology

Psittacine birds are an important source of human infection with *C. psittaci*, but other birds including turkeys, ducks and pigeons may also cause infection. The term ornithosis is, therefore, more correct.

- Infection is more common in adults, reflecting greater exposure to infected birds
- Transmission is through handling infected ill birds or, less commonly, by inhaling organisms in dried dust from bird droppings. Outbreaks occur rarely
- Transmission between birds, especially when held captive, is common. Hence, it is often recently imported birds that are the source of infection
- 20 per cent of patients with psittacosis give no history of bird contact
- Many infections are mild or subclinical
- Birds with *C. psittaci* infection are unwell, although sometimes the features may only indicate a mild illness
- *Chlamydia psittaci* accounts for <3 per cent of cases of community-acquired pneumonia in the UK; there are approximately 300 cases annually
- The IP is 10 days (range 7–15 days).

Pathology and pathogenesis

The primary site of disease in humans is the lung. The bacterium enters through the respiratory tract and, after a transient bacteraemia, seeds the reticuloendothelial system. In the lung, there is an alveolitis followed by an interstitial exudate. Mucus plugging is a feature of chronic cases. A second bacteraemia coincides with the onset of symptoms. Hepatitis is often observed on biochemical testing, and rarely may be frank and the primary presentation; biopsy shows reactive hepatitis but may also show granulomata.

Clinical features

Characteristics are as follows:

- the onset is abrupt, with fever, rigors, headache, myalgia and a dry cough (occasionally with blood streaking); pleuritic pain is rare
- clinical examination of the chest may be unremarkable despite marked involvement on CXR
- splenomegaly is found in one-third and a rash is rarely observed (Horder's spots)
- abdominal pain, diarrhoea, jaundice, pharyngitis and a mild encephalopathy are occasionally seen
- there is a relative bradycardia.

Complications

- Respiratory failure
- Hepatitis, encephalopathy, glomerulonephritis, myocarditis, endocarditis.

Diagnosis

Psittacosis is suggested by:

- a CXR showing soft, patchy infiltrates radiating out from the hila and involving the lower zones
- a normal WCCt and ESR
- mildly deranged LFTs.

Psittacosis is confirmed by:

- a four-fold rise or fall in antibodies or a single high titre by complement fixation. Distinction from *C. pneumoniae* requires a type-specific immunofluorescence assay. Prompt treatment can delay and blunt the antibody response.

Treatment

The treatment is tetracycline; erythromycin is also effective and is the drug of choice in children. Therapy should be continued for 2 weeks. Rifampicin, ciprofloxacin and the new macrolides (e.g. clarithromycin) have also been used successfully but clinical experience is less.

Prevention

- Quarantining and prophylactic administration of tetracycline-medicated seed to imported birds
- Tetracycline supplemented to animal feeds can prevent chlamydial infections in poultry flocks but may create wider problems with antibiotic-resistant *Salmonella*
- There is no vaccine.

Prognosis

Mortality is 1 per cent of cases, but in severe cases complicated by renal insufficiency it may rise to 20 per cent.

Q fever

Epidemiology

Q fever is caused by the rickettsia *C. burnetii*. It is a worldwide zoonoses affecting wild animals as well as domestic livestock. Infection is endemic in cattle, sheep and goats, where most human infection is acquired.

- Infection is largely subclinical in animals
- Transmission is by aerosols or environments (e.g. dust) contaminated by parturition products. Rarely, contaminated raw milk may be responsible

- It is the cause of <1 per cent of community-acquired pneumonia, and an occasional cause of acute hepatitis
- It is an occupational hazard for abattoir workers, veterinarians, etc.; outbreaks have occasionally been reported in these groups and in others exposed to dust from animals
- The IP is 14–28 days.

Pathology and pathogenesis

After inhalation, the organism multiplies in the lung, followed by bacteraemia which coincides with the onset of symptoms. A lymphocytic interstitial pneumonitis occurs, often accompanied by a mild hepatitis. Liver (and bone-marrow) biopsy may show 'doughnut' granulomata with a fatty fibromatous centre. Rarely, patients may present with chronic infection, usually affecting the aortic valve, and with typical clinical and pathological features of endocarditis. Chronic liver disease leading to cirrhosis has also been described.

Clinical features

- The illness starts abruptly with fever, headache, myalgia and occasionally cough
- Clinical or radiological evidence of pneumonia develops in 10–30 per cent of patients. Clinical signs may be few despite quite extensive X-ray abnormalities
- Hepatosplenomegaly is found in 50 per cent of cases.

Complications

- Hepatitis
- Chronic infection: endocarditis, cirrhosis (acute infection nearly always unrecognised).

Diagnosis

- Demonstration of a four-fold rise or fall or single high titre of phase 2 antibodies by complement fixation. Phase 1 antibody is absent in acute infection
- Chronic Q fever should be suspected in all culture-negative cases of endocarditis; both phase 1 and 2 antibodies are detectable
- Phase 1 and 2 antibodies have been found in individuals with no evidence of cardiac involvement, and it is likely in these persons that chronic liver infection is present.

Treatment

The treatment of choice is tetracycline, but response may be slow. Erythromycin, ciprofloxacin, rifampicin and the newer macrolides (e.g.

clarithromycin) are also active. Treatment of endocarditis requires prolonged therapy, preferably with a combination of antibiotics including tetracycline for the first fortnight. Management of endocarditis is dealt with in more detail in Chapter 8.

Prevention

- Pasteurisation of raw milk
- No vaccine is available.

Prognosis

In acute Q fever, fatalities are exceptionally rare. In Q fever endocarditis, mortality is 10–15 per cent of cases.

Mycoplasma pneumonia

Epidemiology

Mycoplasma pneumoniae is a common pathogen of worldwide distribution resulting in both upper (rhinitis, pharyngitis, myringitis, conjunctivitis) and lower (pneumonia, tracheobronchitis) RTI. Infection is usually mild.

- It accounts for approximately 5 per cent of community-acquired pneumonias in the UK; during outbreaks this may treble
- Infection rates are greatest in children and young adults
- Transmission is by droplets and requires close contact. Spread is slow, with weeks between cases, but attack rates at home are high (75 per cent children and 35 per cent adults) with a slight preponderance in males
- Pneumonia is the most commonly recognised form of infection and affects mainly older children and young adults. However, only 5 per cent of infected persons develop pneumonia, nondescript upper RTI being the commonest manifestation, especially in infants and young children
- Infection is endemic in large communities without evidence of seasonal prevalence. However, periodic outbreaks in autumn and early winter occur every 3–4 years
- The IP is 6–23 days.

Pathology and pathogenesis

Attachment to respiratory tract epithelium is by P1-protein binding to neuraminic acid glycoprotein receptors. This is sequentially followed by ciliostasis, deciliation and cytopathic effects. The organism also stimulates both T- and B-cell lymphocyte responses; the relative importance of direct mycoplasmal damage, as opposed to immune mechanisms, has not been determined. I and I erythrocyte antigens serve as receptors for *M. pneumoniae* resulting in cold haemagglutinins and the complication of intravascular haemolysis. Fatal cases show patchy alveolar pneumonitis and peribronchial infiltrate of lymphocytes and plasma cells. Immunity follows primary infection but occasional secondary infections have been described.

Clinical features

Characteristic features include:

- a 2–4 days prodrome of gradual-onset high fever, rigors, malaise, myalgia and headache
- upper RTI (coryza, tonsillopharyngitis, myringitis or conjunctivitis) in 25–50 per cent
- development of a dry cough which is later productive (33 per cent) of mucopurulent or occasionally blood-tinged sputum
- absence of pleuritic chest pain
- unremarkable chest examination despite marked radiological features
- extrapulmonary manifestations. Skin rashes (maculopapular, urticarial, erythema multiforme, erythema nodosum), diarrhoea and vomiting, and arthritis each occur in approximately 15–25 per cent of cases.

Complications

Pulmonary

- Bullous myringitis (visible bleb formation on the tympanic membrane)
- Respiratory failure, adult respiratory distress syndrome (ARDS)
- Pleural effusion (15 per cent)
- Cavitation, pneumatocoeles, bronchiectasis, pulmonary fibrosis (all rare).

Extrapulmonary

- Intravascular haemolysis
- Meningo-encephalitis, transverse myelitis, and cranial and peripheral neuropathies (all rare)
- Hepatitis, glomerulonephritis, pancreatitis, myocarditis and pericarditis (all rare).

Diagnosis

Nonspecific features supporting *M. pneumoniae* include:

- presence of current or antecedent upper RTI and extrapulmonary features
- patchy consolidation of one or other lower lobes (80 per cent), often bilateral (30 per cent)
- presence of cold haemagglutinins (50 per cent)

- total WCCt of $<10 \times 10^9$/litre (90 per cent)
- mildly deranged LFT.

Specific diagnosis is made through one of the following:

- isolation of *M. pneumoniae* from throat or sputum culture using selective liquid media
- detection of nucleic acid in tissue or sputum using DNA probes
- four-fold rise or fall or a single high antibody titre by complement fixation or enzyme-linked immunoadsorbent assay (ELISA).

Treatment

Both erythromycin (or clarithromycin) and tetracycline are effective. Erythromycin tends to be preferred because of better activity against *Legionella* and the problem with tetracycline of staining children's teeth. Therapy should be for 2 weeks. The clinical response is not rapid.

Prevention

There is no vaccine for mycoplasmal infection. Isolation in hospital is advised.

Prognosis

Mycoplasma pneumoniae infection is usually mild and self-limiting. Where pneumonia occurs it tends to be a benign illness. During outbreaks, occasional fatalities are always reported but this is a rare event.

Legionnaires' disease

Epidemiology

Legionella pneumophila was first identified in 1976 and was subsequently identified as the cause of several earlier outbreaks of pneumonia (Legionnaires' disease) and influenza-like illness (Pontiac fever). Over 30 other species of *Legionella* have been identified, of which 16 can cause a similar respiratory illness when it is termed legionellosis. *Legionella pneumophila* is:

- a widespread, naturally occurring aquatic organism. Survival has been associated with blue-green algae and free-living acanthamoeba
- transmitted via airborne water droplets; possible sources include air-conditioning cooling towers, hot-water systems, humidifiers and showers. Case-to-case transmission does not occur
- responsible for 5 per cent of community-acquired pneumonias; there are approximately 250 cases annually in the UK
- divided into 14 serogroups, of which serogroup 1 is most important

- very occasionally responsible for infection at other sites (wound infection, endocarditis).

Legionnaires' disease:

- is most frequently seen in middle-aged males
- results in subclinical infection infrequently, and infection in childhood very rarely
- may occur sporadically or in outbreaks, often centred on institutions such as hotels or hospitals. The attack rate is <5 per cent in outbreaks
- is contracted abroad in 50 per cent of cases
- is most common in patients with diabetes, malignancy, undergoing dialysis, on immunosuppressives, and in heavy smokers and drinkers
- has an IP of 2–10 days.

Pontiac fever:

- is caused by a hypersensitivity reaction to *L. pneumophila*. The organism is never isolated from cases and diagnosis is by serology
- has an attack rate of 90 per cent in outbreaks
- has an extremely good prognosis without specific therapy
- has an IP of 1–2 days.

Pathology and pathogenesis

Legionella pneumophila causes alveolar and interstitial inflammation with the development of patchy or widespread consolidation in the lung. It is an intracellular pathogen, parasitising macrophages where it multiplies and evades host defences. Successful resolution of infection depends primarily upon cell-mediated immunity.

Clinical features

Distinction from other causes of community-acquired pneumonia is difficult. In favour of Legionnaires' disease are the following:

- severe influenza-like symptoms (80 per cent) with fever, rigors, myalgia and headache
- prominence of extrapulmonary symptoms (35 per cent), such as diarrhoea, abdominal pain and confusion
- the later development of chest symptoms with dry cough and dyspnoea. Haemoptysis, chest pain and a productive cough may follow on later
- on examination, systemic toxicity, relative bradycardia, tachypnoea and crepitations on auscultation.

Complications

- Acute respiratory failure
- Pleural effusion/empyema (5 per cent), cavitation (rare)
- Residual fibrosis

- Renal failure
- Cranial and peripheral nerve palsies, ataxia.

Diagnosis

Nonspecific features supporting *L. pneumophila* include:

- limited chest signs with extensive involvement on CXR
- patchy consolidation, often bilateral (30 per cent) and more frequent involvement of the lower zones
- hyponatraemia, hypophosphataemia, and modest elevations of transaminases, alkaline phosphatase, bilirubin, urea and creatinine
- total WCCt of $<15 \times 10^9$/litre and lymphopenia
- microscopic haematuria and proteinuria (30 per cent).

Specific diagnosis is made through one of the following:

- sputum culture (positive in 15 per cent)
- antigen detection in tissue, sputum or urine
- four-fold rise or fall or single high antibody titre by immunofluorescence (IFAT) or agglutination (RMAT, Latex). 25 per cent of patients only seroconvert 4–8 weeks after onset.

Treatment

- Erythromycin is the antibiotic of choice (clarithromycin is an alternative). In severe cases, either rifampicin or ciprofloxacin should be added. Treatment should be continued for 2 weeks
- Intensive care may be needed.

Prevention

With an established outbreak, it is important to identify the source or site of water exposure and, if possible, close it down and screen at-risk patients (e.g. those in a dialysis unit). Treatment of contaminated water supply by hyperchlorination or maintenance of water storage temperatures at $>60°C$ and outlet temperatures at $>50°C$ will reduce the risk of institution-associated infection.

Prognosis

The mortality rate in epidemics of *L. pneumophila* pneumonia is between 10 and 25 per cent of cases.

Pneumococcal pneumonia

Epidemiology

Streptococcus pneumoniae is responsible for 20–30 per cent of cases of community-acquired pneumonia (incidence: 1 case/1000 adults in the

UK), is the second most frequent pathogen in bacterial meningitis, and is a major aetiological agent in sinusitis, otitis media and infective exacerbations of chronic chest disease. Pneumococci are:

- commonly carried asymptomatically in children (30 per cent) and adults (15 per cent). Carriage is higher where there is overcrowding or a closed group
- transmitted by droplet infection; this is augmented by overcrowding and coexistent viral upper RTI. Pharyngeal colonisation usually precedes pneumonia.

Pneumococcal pneumonia:

- is more frequent in patients with HIV infection, the nephrotic syndrome, chronic lung, heart and kidney disease, splenic dysfunction or post-splenectomy, alcoholic liver disease or cirrhosis, and patients with certain malignancies (e.g. myeloma, lymphoma) and severe immunosuppression
- is commoner in males, persons aged >40 years, during winter and in closed groups – in this context outbreaks occasionally occur
- is a common complication of respiratory viral infections, particularly influenza
- has an IP of 1–3 days.

Pathology and pathogenesis

Pneumococci are capsulate (imparting virulence) and the antigenic differences between the capsules form the basis of the serotypic classification. There are over 80 serotypes and certain serotypes are more commonly associated with clinical disease; polyvalent pneumococcal vaccine contains 23 of the commonest serotypes isolated. Following inhalation, *S. pneumoniae* rapidly multiply in the alveolar spaces. Subsequently, the classical pathological progression of congestion, through red hepatisation to grey hepatisation and finally to resolution, occurs. The inflammation is purely alveolar and there is no development of necrosis or post-pneumococcal fibrosis.

Clinical features

In acute pneumococcal pneumonia:

- the onset is abrupt, with a rigor followed by a high fever and general prostration
- a dry cough, which later becomes productive of rusty sputum, and chest pain develop in 75 per cent of patients
- examination reveals toxicity, tachypnoea and features of consolidation.

Complications

Local

- Pleural effusion (10 per cent)
- Empyema (< 1 per cent)
- Pericarditis.

Empyema should be suspected where persistent fever and leucocytosis occur, despite adequate antibiotic therapy; aspiration and/or drainage is indicated. Pericarditis results from contiguous spread.

Systemic

- Bacteraemia (occurs in 25 per cent and doubles mortality)
- Meningitis, arthritis, endocarditis (rare).

In patients without a functioning spleen, a fulminant course may be observed with death in 24 hours.

Diagnosis

Nonspecific features supporting *S. pneumoniae* include:

- presence of chest pain and few extrapulmonary features
- leucocytosis $>15 \times 10^9$/litre
- radiological evidence of well-defined area(s) of consolidation (Fig. 7.2)
- sputum Gram stain showing Gram-positive diplococci.

Confirmation of infection may come from:

- isolation of a pure and heavy growth of *S. pneumoniae* from sputum culture
- recovery of *S. pneumoniae* from blood culture or pleural fluid

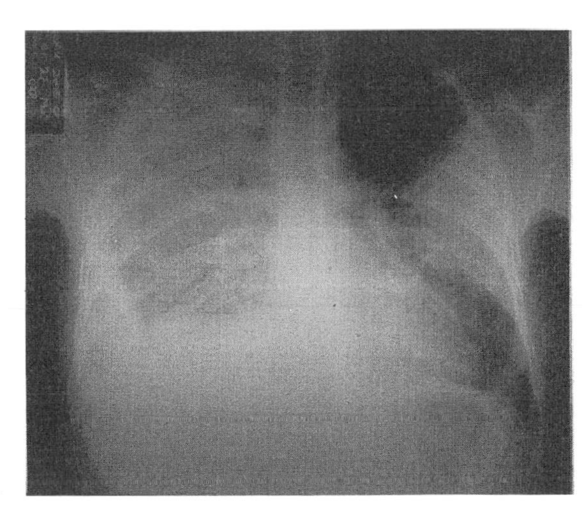

Fig. 7.2. Severe pneumococcal pneumonia.

• pneumococcal antigen detection in sputum, pleural fluid, urine and/or serum.

Treatment

The recognition of penicillin-resistant pneumococci has altered management in many areas of the world. These strains are characterised as follows:

• they are associated with prior antibiotic use, the extremes of age and hospital acquisition

• they account for >30 per cent of all isolates from clinically important infections in certain areas of North and South America, Spain and South Africa

• they can be split into those with intermediate sensitivity to penicillin (high-dose benzylpenicillin effective except in meningitis) and fully resistant strains (benzylpenicillin completely ineffective)

• they may be associated with high-level resistance to other antibiotics.

Where penicillin resistance does not occur, the treatment is benzylpenicillin. Where intermediately resistant strains are common but fully resistant strains rare, cefotaxime should be used; where fully resistant strains are endemic, a combination of cefotaxime and vancomycin is indicated. When the sensitivity profile of the pneumococcus is known, treatment can be changed accordingly; benzylpenicillin is still the drug of choice for penicillin-sensitive strains.

Prevention

• Immunisation with pneumococcal vaccine containing capsular polysaccharide of the commonest 23 serotypes gives 60–70 per cent protection. It is indicated for the at-risk groups mentioned above. Immunity is long lasting. The currently available unconjugated vaccine is poorly immunogenic in children <2 years of age

• Chemoprophylaxis with oral penicillin is indicated in children with sickle cell anaemia.

Prognosis

Pneumococcal pneumonia is a leading cause of death. The conditions that predispose to infection also increase the likelihood of severe and fatal disease. In community-acquired pneumonia, the presence of two of the following parameters is associated with a 21-fold increased mortality:

• respiratory rate >30/min

• diastolic blood pressure <60 mmHg

• blood urea >7 mmol/litre.

Whooping cough

Epidemiology

Whooping cough is a highly infectious disease caused by *Bordetella pertussis*. The characteristic whoop is a result of a sharp indrawing of breath which follows a paroxysm of coughing. Several viruses (particularly adenovirus), *M. pneumoniae*, *C. trachomatis* and *B. parapertussis* can all produce a similar but milder illness. Whooping cough:

- is a major cause of childhood illness (600,000 deaths annually) worldwide. Most infections occur in unimmunised infants
- is a disease of young children; 50 per cent of cases occur in those <1 year of age, when the mortality is highest: there is no effective transmitted maternal immunity. Adults can get the disease when the illness is milder; they may be the source for children
- is an endemic disease with epidemics every 3–5 years
- is readily transmissible from clinical cases by aerosol, with attack rates of over 50 per cent
- occurs more frequently and seriously in females
- confers good immunity after a single attack; second attacks are rare
- is infectious from the onset of catarrhal symptoms for up to 1 month
- has an IP of 7–14 days.

Pathology and pathogenesis

The primary pathology is in the lungs, with systemic manifestations being secondary to toxaemia and respiratory complications. Pulmonary atelectasis is common and is due partly to bronchial blockage by viscid mucus and partly to bronchial and peribronchial inflammation. The extent of collapse varies from small subsegmental areas to a whole lobe. In fatal cases, the major pathological feature is haemorrhage, which can be explained by the raised pressure caused by paroxysmal coughing. *Bordetella pertussis* produces a number of biologically active substances:

- the filamentous haemagglutinin and the agglutinogens (important in attaching to ciliated respiratory epithelium)
- tracheal cytotoxin
- dermonecrotic and pertussis toxins (causing ciliostasis and local cell damage)
- pertussis and adenylate cyclase toxins (interfering with phagocyte function)
- pertussis toxin (causing the systemic manifestations of disease). This toxin has a typical A/B unit structure with binding and active toxin

components and is directed against guanine nucleotide binding (G) proteins.

Clinical features

The early clinical features, with coryza, mild fever and dry cough, are nonspecific but the child is highly infectious. However, over the next 1–2 weeks other signs will be seen:

- the cough becomes progressively more severe
- paroxysms develop with the characteristic whoop on indrawing of breath. The whoop is uncommon in infants and adults, and may be absent in those with partial immunity
- cough and whoop become more frequent, especially at night, occurring up to 50 times/day and usually terminated by vomiting. Paroxysms can be triggered by handling or examining the child, or by exposure to cold air
- cyanosis during paroxysms commonly occurs.

Apnoeic attacks may be the only symptoms in the very young, as may chronic cough in the adult. After 2–4 weeks, the paroxysms become gradually less frequent, but may continue to occur especially at night for up to 6 months. Any minor intercurrent infection may re-trigger paroxysms.

Complications

Respiratory tract

- Pulmonary collapse (particularly the lower lobes and the right middle lobe)
- Secondary bacterial pneumonia.

Other complications

- The increased pressures associated with paroxysmal coughing can result in subconjunctival haemorrhages, epistaxis, facial and truncal petechiae, subcutaneous emphysema, pneumothorax, abdominal hernias and rectal prolapse. Ulceration of the lingual frenulum due to repeated tongue movement over the lower incisors occasionally occurs
- Convulsions complicate a minority (2 per cent) and may result from cerebral haemorrhage (rare), anoxia or neurotoxic properties of pertussis toxin
- Vomiting can lead to weight loss and malnutrition.

Diagnosis

During the early catarrhal phase, the diagnosis can only be suspected from a history of contact. With the development of typical paroxysms and whooping, the diagnosis becomes clinical. Nonspecific features supporting *B. pertussis* include:

• a persistent paroxysmal cough, worse at night, and accompanied by vomiting
• the appearance of the whoop
• leucocytosis (>20 × 10^9/litre) with lymphocytosis (>80 per cent) (not usually seen in infancy)
• pulmonary collapse or consolidation (20 per cent).

Confirmation can be obtained from:
• per nasal swab culture on selective charcoal-based blood or Bordet–Gengou media
• direct pertussis antigen detection on nasopharyngeal aspirates.

Treatment

Most children under 1 year of age, and older children with complications, will need to be hospitalised.
• The older child without complications is allowed to be up and about. Support and reassurance are needed. Meals are kept small and frequent and, if vomiting is persistent, should be given after a paroxysm
• Erythromycin for 2 weeks eliminates *B. pertussis*.

For those admitted to hospital, the nursing management is vital.
• All cases should be isolated where possible
• Oxygen saturation should be monitored and oxygen should be given for recurrent or sustained hypoxaemia
• Gentle nasal suction should be performed to remove secretions
• Adequate hydration and nutrition need to be maintained, intravenously if necessary
• Disturbance of any sort should be kept to a minimum to reduce paroxysms
• Secondary bacterial pneumonia requires parenteral antibiotic treatment (e.g. cefuroxime)
• Major pulmonary collapse requires prolonged and intensive physiotherapy.

Steroids, β-agonists and sodium cromoglycate have no effect on the paroxysms.

Prevention

• Control of the disease is mainly by immunisation. A full course of killed whole-cell vaccine (three doses at 2, 3 and 4 months) confers protection in over 80 per cent of subjects for a decade. Less severe disease occurs in those not fully protected. Vaccines must contain all three major agglutinogens
• There is still no evidence of a causal relation between pertussis vaccine and permanent neurological illness, and confidence in the vaccine has improved (the vaccination rate has increased from 20 per cent in 1980 to

87 per cent presently in the UK). Nevertheless, acellular vaccines which have fewer side effects have been developed. In the developing world only one-third of infants have been immunised

- Erythromycin given for 14 days to a contact will attenuate the illness if the child is in the catarrhal stage
- Erythromycin should be given to non- or partially-immunised infant contacts.

Prognosis

Good nursing and medical practice reduce the complications, but morbidity is high. In England in 1982, more than 65,000 cases were reported, with 14 deaths (0.02 per cent).

Pulmonary tuberculosis

Epidemiology

Tuberculosis is the most common infectious disease in the world. A falling incidence has been reversed with increases in both developed and developing nations since the mid-1980s. Much of the new epidemic is fuelled by HIV; the World Health Organisation predict that HIV-attributable tuberculosis deaths will increase from 4 to 14 per cent of overall tuberculosis mortality by the year 2000. *Mycobacterium tuberculosis*:

- belongs to a complex of organisms (*M. tuberculosis* complex) with *M. bovis* (reservoir cattle) and *M. africanum* (reservoir humans) which cause clinical tuberculosis
- infects 8 million new cases every year and causes 2.5 million deaths. This is projected to increase to 14 and 3.5 million, respectively, by the year 2000
- is increasing as a result of many factors: HIV co-infection, increasing survival of the elderly (infected as children and now reactivating), worsening social deprivation, immigration from countries with high prevalence, and increasing drug resistance
- is much commoner in immigrants from countries with a high prevalence of the disease; in the UK, the incidence is 25 times greater in persons from the Indian subcontinent
- causes pulmonary (75 per cent) and extrapulmonary disease (25 per cent)
- can result in occasional large outbreaks in institutions (schools, paediatric wards)
- is transmitted by aerosol from smear-positive pulmonary cases. 10 per cent of contacts develop primary tuberculosis which can be detected by

a positive tuberculin skin test (Heaf or Mantoux): most cases of primary disease occur in young children. 5 per cent of those infected develop progressive primary infection; another 5 per cent of persons will reactivate in later life (post-primary tuberculosis). Primary disease is rarely infectious whereas post-primary tuberculosis is usually pulmonary and ZN-positive. In adults, tuberculosis is usually a result of reactivation (90 per cent)

- in HIV-positive persons results in a greater likelihood of contracting disease if exposed, of developing progressive primary disease, of developing extrapulmonary and atypical pulmonary manifestations and of reacting to standard drugs
- can be resistant to one or more of the anti-tuberculosis drugs. Primary and secondary (past treatment) resistance is uncommon in the UK (<5 per cent) but is frequent in many developing countries. Multidrug resistant strains (MDRTB) have recently been identified in HIV care facilities
- has an IP for primary infection of 4–16 weeks.

Pathology and pathogenesis

Following inhalation of *M. tuberculosis*, the disease progresses as follows:

- a small subpleural lesion, termed a Ghon focus, develops
- infection then spreads to the hilar and mediastinal lymph glands to produce a primary complex. These enlarge with inflammatory granulomatous reaction, and may caseate. In 95 per cent of cases, the primary complex heals spontaneously in 1–2 months, sometimes with calcification, and the individual becomes tuberculin skin test positive
- in a minority, the infection spreads from the primary complex, locally to a bronchus causing a segmental lung lesion (endobronchial, bronchopneumonia or collapse), via the lymphatics to the pleura causing a pleural effusion, or via the bloodstream to cause disseminated lesions (miliary when in the lung)
- in some, disease then progresses with the development of miliary or meningeal tuberculosis. In others, dormant foci are created in the bone, lungs, kidneys, etc., which reactivate in later life.

The virulence factors of *M. tuberculosis* have not been fully elucidated. The organism is versatile, with the ability to multiply rapidly outside cells within cavities, survive inside macrophages and to prevent fusion of the lysosome and phagosome, and to survive in a relatively inactive state with only infrequent bursts of division.

Clinical features

- The primary pulmonary complex is often asymptomatic or subclinical.

Endobronchial tuberculosis may result in wheezing and cough. Collapse from obstruction or tubercular bronchopneumonia is usually associated with constitutional symptoms, cough and sputum

• Miliary tuberculosis is a severe infection, often diagnosed late. The patient has fever, tiredness, sweating, often with a dry cough and mild dyspnoea, anorexia and weight loss. Hepatosplenomegaly may be present; auscultation of the chest may be normal. Choroidal tubercles may be seen. The CXR shows fine 'millet-seed' appearances through both lungs

• Post-primary adult pulmonary tuberculosis is the most frequent presentation. Cough, haemoptysis, dyspnoea, anorexia, weight loss associated with fevers and sweats is typical. The history is subacute (4–6 weeks) in the majority. Auscultation of the chest usually reveals localised signs.

Complications

Local

- Respiratory failure, ARDS
- Haemoptysis (occasionally massive)
- Lung fibrosis, cor pulmonale
- Pleural effusion, empyema
- Pericarditis (effusion and constriction)
- Atypical mycobacterial colonisation (e.g. *M. malmoense*)
- Aspergilloma.

Systemic

- Meningitis
- Spread to any organ.

Diagnosis

Nonspecific features supporting tuberculosis include:

- subacute illness with fever, weight loss and cough
- haemoptysis
- CXR showing a cavity (usually apical), or bilateral disease (Fig. 7.3)
- raised ESR, mild normocytic normochromic anaemia, normal WCCt, mildly deranged LFTs
- history of foreign residence, recent immigration or HIV infection.

Confirmation of tuberculosis is by:

- ZN microscopy
- culture of *M. tuberculosis* on Lowenstein–Jensen or Bactec media
- DNA amplification (PCR).

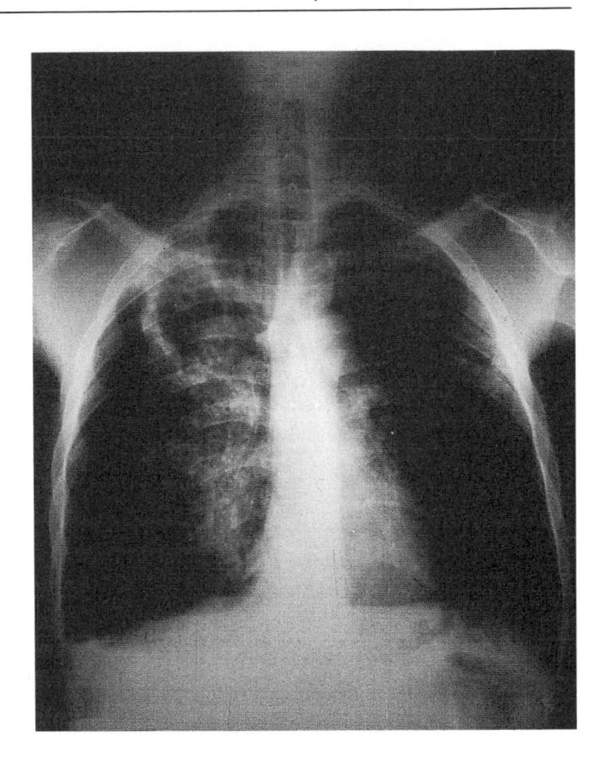

Fig. 7.3. Ziehl–Nielsen positive pulmonary tuberculosis.

Treatment

- Tuberculosis is treated with multiple drug therapy to inhibit the multiplication of low-level resistant *Mycobacterium*
- Standard therapy is isoniazid for 6 months, rifampicin for 6 months and pyrazinamide for 2 months
- If resistance is suspected (e.g. past history of treatment, immigration), ethambutol for 2 months should be included until the results of sensitivity are known
- Treatment should be supervised as closely as possible for the first 2 weeks, partly to see that compliance is satisfactory and partly to be alert to drug reactions. After the first 2 weeks of treatment, patients can be regarded as non-infectious
- The most important drug reactions are hepatitis (isoniazid, rifampicin, pyrazinamide), optic neuritis (ethambutol), psychosis (isoniazid), peripheral neuropathy (isoniazid) and skin rashes (streptomycin and thiacetazone in HIV patients)
- Surgery is occasionally needed for localised areas of bleeding

• Corticosteroids are beneficial in pericardial and pleural disease and anecdotally in preventing the development of a Herxheimer-like reaction in severe pulmonary disease
• Multidrug-resistant *M. tuberculosis* (MDRTB) is resistant to rifampicin and isoniazid, with some strains being resistant to eight first- and second-line drugs. They have developed in areas with high prevalence of HIV and are often nosocomial.

Prevention

The best protection against tuberculosis is the efficient diagnosis and treatment of people with active infections. Tuberculosis of all forms is a notifiable disease (see p. 5).
• Household and other similarly close contacts of patients with pulmonary tuberculosis are screened by tuberculin skin test (usually Heaf) and CXR
• The tuberculin skin test involves intradermal injection of inactivated *M. tuberculosis* and is read at 3–7 days. A positive reaction is graded (e.g. Heaf grades 0–4) dependent on the degree of induration. The test is used to assess whether a person has acquired *M. tuberculosis* following exposure, and is useful in persons not immunised with BCG. It is also used pre-immunisation with BCG to judge whether or not persons have had previous primary tuberculosis. Interpretation is more difficult in BCG-vaccinated persons since a mild positive reaction is to be expected
• Depending on their age, BCG history, Heaf result and CXR, contacts are given primary prophylaxis (isoniazid and rifampicin or isoniazid alone), BCG immunisation, reviewed annually or discharged. Occasionally contacts are found to have active disease in which case they are treated
• BCG is used in some countries as a protective measure for mycobacterial infections. It gives approximately 80 per cent protection for 20 years. Some countries do not use it because they value the diagnostic sensitivity of the tuberculin skin test as a measure of recent primary infection, which is lost after immunisation. Occasional complications include local BCG abscesses, and disseminated BCG infection in immunocompromised persons.

Prognosis

With isoniazid, rifampicin and pyrazinamide chemotherapy for 6 months, cure is to be expected and the likelihood of relapse at 5 years is <2 per cent. In countries where these drugs are unavailable, a combination of isoniazid, streptomycin (for the first 2 months) and thiacetazone needs to be continued for a mimimum of 1 year to obtain equivalent results. In

HIV-associated tuberculosis, mortality is increased, but mainly as a result of superimposed bacterial infections.

Cross-references

Tuberculous lymphadenitis (see p. 64).
Tuberculous meningitis (see p. 125).
Abdominal tuberculosis (see p. 207).

CHAPTER 8

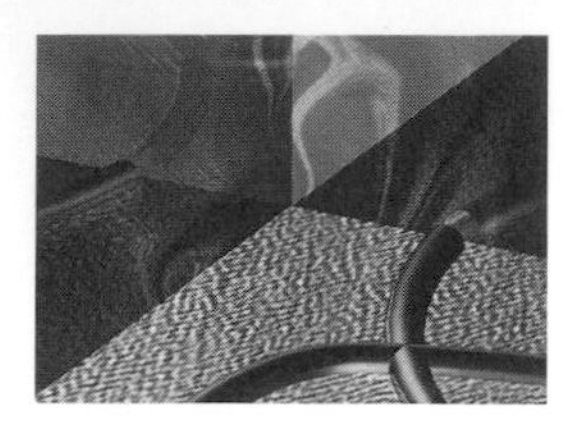

Infections of the Cardiovascular System

CLINICAL SYNDROMES

Endocarditis, myocarditis and pericarditis

Major presenting features: *breathlessness, chest pain, fever.*

Causes

Endocarditis

Predisposing:

• rheumatic heart disease, congenital (e.g. bicuspid aortic valve, ventricular septal defect (VSD), patent ductus arteriosus (PDA), degenerative valvular disease, mitral valve prolapse, prosthetic valve)

• intravenous (IV) drug abuse, preceding dental or other instrumentation.

Microbiological — native valve (80 per cent):

• STREPTOCOCCUS VIRIDANS (*S. sanguis, S. mitior, S. bovis, S. milleri, S. mutans, S. salivarius*), group C and G β-haemolytic streptococci, B_6-dependent streptococci

• STAPHYLOCOCCUS AUREUS, ENTEROCOCCUS FAECALIS, *Staphylococcus epidermidis*, 'coliforms'

• *Coxiella burnetii, Chlamydia, Legionella, Mycoplasma*

• *Streptococcus pyogenes* (rheumatic fever).

Microbiological — prosthetic valve (20 per cent):

• *Staphylococcus epidermidis, S. aureus*, diphtheroids, *E. faecalis*, 'coliforms', *Candida.*

Microbiological — IV drug abuser:

• *Staphylococcus aureus, Pseudomonas aeruginosa*, 'coliforms', *Candida, E. faecalis, S. viridans.*

Myocarditis

- COXSACKIE VIRUS, ECHO VIRUS
- Cytomegalovirus (CMV), mumps, Epstein–Barr virus (EBV), human immunodeficiency virus (HIV), influenza, adenovirus
- *Chlamydia*, *Coxiella*, *M. pneumoniae*
- *Streptococcus pyogenes* (rheumatic fever), *Corynebacterium diphtheriae*
- Chagas' disease, trichinosis, toxoplasmosis.

Pericarditis

- COXSACKIE VIRUS, ECHO VIRUS, mumps, EBV, influenza, adenovirus
- *Staphylococcus aureus*, *S. pyogenes*, *S. pneumoniae*, 'coliforms'
- *Streptococcus pyogenes* (rheumatic fever)
- *Mycobacterium tuberculosis*, histoplasmosis, *Entamoeba histolytica*.

Distinguishing features

	Acute viral myocarditis	*Bacterial endocarditis*	*Acute viral pericarditis*
IP	3–10 days	2–6 weeks	3–10 days
Season	summer/autumn	all year	summer/autumn
Chest pain	occasional	uncommon	common
Dyspnoea	prominent	rare	rare
Fever	absent/low	low grade	absent/low
Tachycardia (> systemic illness)	present	absent	absent
Heart murmur	occasional	present, new or changing	absent
Gallop rhythm	present	absent	absent
Signs of heart failure			
left	present	uncommon	absent
right	present	uncommon	if effusion
Arrythmias	common	rare	rare
Pericardial rub	absent	absent	present
Immunological lesions*	absent	often present	absent
Splenomegaly	rare	50%	rare
Clubbing	absent	20%	absent
Emboli	absent	occasional	absent
Blood cultures	negative	positive	negative

*Osler's nodes, splinter haemorrhages, Janeway lesions, Roth spots

	Acute viral myocarditis	*Bacterial endocarditis*	*Acute viral pericarditis*
Microscopic haematuria	absent	50%	absent
Anaemia	rare	common	rare
CXR			
cardiomegaly	often	uncommon	if effusion
LVF	often	uncommon	absent
ECG			
T waves	inverted	normal	concave rise
QRS complex	may be widened	normal	normal
Echocardiogram	poor left ventricular function	vegetations in half	normal or effusion

Complications

Endocarditis: heart failure (acute/chronic), major embolus (carotid, coronary, limb), mycotic aneurysm, cardiac abscess, encephalopathy, glomerulonephritis, pericarditis, R–L fistulas

Myocarditis: heart failure (acute), arrythmias

Pericarditis: effusion, tamponade, constriction (tuberculosis).

Investigations

For all three infections

- Full blood count (FBCt), differential white cell count (WCCt) and erythrocyte sedimentation rate (ESR)
- Biochemical profile (liver function tests (LFTs), albumin and urea/creatinine)
- Chest X-ray (CXR)
- Cardiac enzymes to exclude myocardial infarction (MI)
- Electrocardiogram (ECG)
- Echocardiogram.

Endocarditis

- Blood cultures (× three)
- Midstream urine (MSU):
  - blood and protein
  - microscopy and culture
- Serology:
  - *C. burnetii* (phase 2)
  - full screen of viral and other atypical agents (if negative cultures)

- Minimum inhibitory/bactericidal concentrations (MICs/MBCs) (to confirm antibiotic sensitivities/choice)
- Serum back-titrations (to monitor therapy).

Myocarditis and pericarditis

- Viral cultures: throat swab and faeces
- Serology
  - antistreptolysin O (ASO) titre (rheumatic fever)
  - full screen of viral and other atypical agents
- Pericardial aspirate (microscopy, Gram and Ziehl–Nielsen (ZN) stains, culture and cytology)
- Biopsy:
  - endomyocardial to confirm and identify cause of myocarditis
  - pericardial to identify cause of pericarditis with effusion.

Differential diagnoses

Endocarditis: non-bacterial vegetations as seen in malignancy, systemic lupus erythematosus (SLE) (Liebman–Sacks), chronic wasting illnesses, uraemia; seronegative athritides (e.g. ankylosing spondylitis, Reiter's syndrome), tertiary syphilis

Myocarditis: sarcoidosis, cardiomyopathy, uraemia, radiation, drugs (doxorubicin, emetine), lead poisoning, toxaemia, MI

Pericarditis: MI, Dressler's syndrome, uraemia, hypothyroidism, connective tissue diseases (Still's disease, SLE, rheumatoid disease), lymphoma/leukaemia, carcinoma, post-surgery, trauma, post-radiation.

Treatment

All three infections need bed rest, anti-arrythmic and anti-heart failure drugs as indicated.

Endocarditis

- IV antibiotics:
  - *S. viridans:* benzylpenicillin
  - *S. aureus:* flucloxacillin
  - *E. faecalis:* ampicillin
  - with initial gentamicin for 2 weeks followed by oral antibiotics for 4 weeks
- Surgery, in the case of acute valvular incompetence, prosthetic valve disease or cardiac abscess.

Myocarditis

- Antimicrobial agent(s) if specific cause identified
- Immunosuppressives if severe and presumed of viral aetiology.

Pericarditis

- Antimicrobial agent(s) if specific cause identified (e.g. rifampicin, isoniazid and pyrazinamide for tubercular pericarditis)
- Aspirin or other non-steroidal anti-inflammatory drug (NSAID) (steroids are sometimes used) if presumed viral
- Aspiration if large effusion or tamponade present or imminent.

Rheumatic fever

- Aspirin (corticosteroids if severe).

Prevention

Endocarditis

- Antibiotic prophylaxis should be given during procedures (dental, surgical) for patients at risk (predisposing causes above) (see p. 109).

SPECIFIC INFECTIONS

Endocarditis

Epidemiology

Infective endocarditis results from an infection of the cardiac valves. Although usually subacute and due to a bacterium, it may be acute and have a non-bacterial aetiology, hence the preference for the term 'infective endocarditis' rather than 'subacute bacterial endocarditis'. Infective endocarditis:

- has an incidence of 6–7/100,000 cases, resulting in 3000–4000 cases/year in the UK
- usually involves a previously abnormal value — rheumatic heart disease, congenital abnormalities (e.g. bicuspid aortic valve, VSD), degenerative disease (aortic and mitral valve sclerosis), mitral valve prolapse and prosthetic valves
- is increasing in frequency in developed countries. This is as a result of an increasing elderly population with degenerative valvular disease, more invasive intravascular procedures, longer survival of children with congenital heart disease, increasing IV drug abuse and increasing numbers of patients with prosthetic valves

- occurs mainly in the elderly (50 per cent >50 years of age)
- most frequently affects the mitral and aortic valves, either alone (45 per cent and 30 per cent, respectively) or together (20 per cent)
- may affect previously normal valves
- patients have a history of antecedent dental treatment in 15 per cent of cases
- has an IP of 2–6 weeks.

Pathology and pathogenesis

Classically, platelets and fibrin initiate a vegetation on endothelial breaks which are then colonised by a circulating organism. The bacteraemia follows on from trauma to a colonised mucosal surface, usually the mouth. The vegetation develops at sites lateral to maximum turbulence (small defects produce more turbulence than large ones) and downstream to the defect (e.g. atrial side if atrioventricular valve and ventricular side if semilunar valve). Hence, patients with mitral stenosis or an atrial septal defect (ASD) rarely develop endocarditis, whereas those with aortic incompetence or a VSD are prone to. Where endocarditis develops on previously normal valves, the organism tends to be more virulent and destruction of the valve is rapid and may lead to perforation of the cusp or rupture of the chorda tendinae or papillary muscle. The clinical features result from embolism and/or immune-complex deposition. Major vessel infarcts, most mycotic aneurysms and microscopic haematuria, arise from embolisation with infarction. Vasculitic skin lesions, splinter haemorrhages, Roth spots (retina), Osler's nodes and glomerulonephritis are predominantly immunological. Circulating immune complexes are seen only where infection has been subacute.

Clinical features

The classic features are:

- fever (usually low grade), drenching sweats, influenza-like symptoms (myalgia, arthralgia, malaise) and weight loss
- clubbing (20 per cent), splenomegaly (50 per cent) and anaemia (80 per cent)
- changing heart murmur
- splinter haemorrhages (60 per cent), vasculitic skin lesions (Janeway spots) (5 per cent), Osler's nodes (10 per cent), Roth spots (5 per cent), mucous membrane petechiae (30 per cent – conjunctivae and palate)
- evidence of major embolism. In right-sided disease, the infarcts are pulmonary.

In acute endocarditis, hectic fever, rigors and toxaemia are frequent, whereas skin features are rare. Endocarditis should always be considered as a possible diagnosis in many presentations, particularly in patients:

- with a pyrexia of unknown origin (PUO) and heart murmur
- with unexplained cardiac failure, embolic episodes or a 'vasculitic' rash
- where *S. viridans* has been isolated from a blood culture
- with a prosthetic valve
- who are IV drug abusers.

Complications

Cardiac

- Cardiac abscess
- Cusp perforation, ruptured chorda tendinae or papillary muscle
- Coronary artery embolus
- Acute or subacute ventricular failure
- Pericarditis.

Systemic

- Embolism (coronary, carotid, splenic, renal or limb)
- Mycotic aneurysm
- Glomerulonephritis (focal, diffuse and membranoproliferative)
- Encephalopathy.

Diagnosis

Nonspecific features supporting a diagnosis of endocarditis include:

- low-grade fever, heart murmur, 'vasculitic' lesions, splenomegaly
- normocytic normochromic anaemia, elevated ESR, microscopic haematuria
- multiple infarcts on CXR (right-sided disease).

Confirmation of endocarditis is by:

- demonstration of vegetations by echocardiography (seen in 50 per cent). The sensitivity can be increased by using a transoesophageal probe
- recovery of an organism from blood cultures:
  (a) in native valve endocarditis:
  - *S. viridans* (40 per cent), *S. aureus* (25 per cent) and *E. faecalis* (20 per cent) in those with subacute disease (who are not IV drug abusers); *S. aureus* in those with acute endocarditis
  - *S. aureus*, *P. aeruginosa*, *Candida* and 'coliforms', or the isolation of $\geqslant 1$ organism from IV drug abusers
  - no organisms are isolated in 10 per cent (culture-negative cases). This may be due to recent antibiotic administration or the presence of pyridoxine-dependent *S. viridans*, *C. burnetii*, fungi or other fastidious micro-organisms

(b) in prosthetic valve endocarditis:

- disease is divided into 'early' (within 60 days) and 'late' infection
- 'early' endocarditis indicates infection acquired at the time of surgery; the causes are *S. epidermidis* (30 per cent), *S. aureus* (20 per cent), Gram-negative bacilli (18 per cent), diphtheroids (10 per cent) and *Candida* (10 per cent)
- 'late' endocarditis has a similar microbiological aetiology to native valve disease.

Treatment

Endocarditis is potentially curable: this is more likely the earlier treatment is initiated. Treatment comprises:

- IV antibiotics:
  - *S. viridans:* benzylpenicillin and gentamicin (2 weeks)
  - *S. aureus:* flucloxacillin and gentamicin (2 weeks)
  - *E. faecalis:* ampicillin and gentamicin (2 weeks minimum)
  - *S. epidermidis:* vancomycin and additional antibiotics depending on sensitivity pattern
- oral antibiotics (4 weeks) after initial IV treatment:
  - *S. viridans:* amoxycillin
  - *S. aureus:* flucloxacillin
  - *E. faecalis:* amoxycillin
- surgery: may be needed for acute valvular incompetence, prosthetic valve disease (to replace prosthesis) and cardiac abscess
- bed rest, anti-arrythmic and anti-heart failure drugs as indicated.

Prevention

- At-risk patients undergoing procedures liable to be complicated by bacteraemia should be given antibiotic prophylaxis. Regimens depend upon the type and site of procedure and whether or not the patient has had a previous attack of endocarditis; has a prosthetic valve; or has received penicillin during the previous month. Recommendations are based on the following antibiotics: amoxycillin (or erythromycin or clindamycin if penicillin-allergic), amoxycillin and gentamicin (genitourinary procedure and other complicated cases) and vancomycin and gentamicin (complicated cases where penicillin-allergic or recently received penicillin)
- Patient education.

Prognosis

In typical subacute endocarditis due to *S. viridans*, mortality is only 5 per cent. In acute endocarditis due to *S. aureus* or prosthetic valve-associated endocarditis, mortality is much higher (early 50 per cent, late 25 per cent).

Rheumatic fever

Epidemiology

Rheumatic fever is a multisystem disease which follows on from an *S. pyogenes* throat infection in the preceding 2–4 weeks. It is characterised by migratory arthritis and pancarditis, and:

- results from tonsillopharyngeal infection with rheumatogenic M-types of *S. pyogenes* (e.g. M-type 18); it is not seen complicating infection at other sites. Incidence is closely linked to that of *S. pyogenes* tonsillitis
- has an incidence of up to 1/1000 in developing countries, 100 times more common than in the West. It is a disease of overcrowding and low socioeconomic status
- is a disease of childhood (5–15 years) when it may be missed, patients later presenting with rheumatic heart disease. Infection does occur in adults but it is rare
- case numbers have been increasing recently in Western countries associated with a return of rheumatogenic M-types
- is more severe, and progresses more rapidly, in developing nations reflecting earlier age of first infection and more recurrences
- has an IP of 2–4 weeks.

Pathology and pathogenesis

The pathognomic lesion of rheumatic fever is Aschoff's nodule, a perivascular aggregate of lymphocytes and plasma cells surrounding a fibrinoid core. The basic disease process is that of an inflammatory vasculitis. In the heart, mitral valvulitis is the most common lesion, associated with small, firmly adherent vegetations. The acute pathological changes are followed by fibrosis and subsequent deformity of the valve cusps and chorda tendinae, leading to valve stenosis and/or incompetence. Rheumatic fever results from an aberrantly hyperreactive immune response to *S. pyogenes* antigens. Both cell-mediated and humoral responses are exaggerated and a genetic susceptibility to the disease is likely.

Clinical features

Because the diagnosis is based on clinical findings, these have been well characterised.

- Arthritis is the most common manifestation (80 per cent of cases) and is migratory, large joint, asymmetrical and non-deforming. Pain and tenderness exceed redness and swelling. As one joint improves, another becomes affected; the whole process lasts 3–6 weeks. The response to aspirin is dramatic
- Carditis occurs in half of cases, and may be asymptomatic. It may lead to subsequent rheumatic heart disease. Mitral and aortic valvulitis, myocarditis and pericarditis may all occur
- Chorea is a late manifestation (2–6 months) and can occur alone or with carditis (which may persist for 6 months), affecting 10 per cent overall. It is more common in girls aged 8–12 years
- Erythema marginatum (occurring in 5 per cent) is a non-pruritic skin rash, appearing as pink macules which clear centrally leaving a rim of red which spreads
- Subcutaneous nodules are round, firm and painless and appear over bony prominences such as the occiput and elbow
- Low-grade fever is usually present.

Complications

- Pancarditis
- Recurrent rheumatic fever
- Rheumatic heart disease.

Rheumatic heart disease is the major complication. The probability of recurrent rheumatic fever after further *S. pyogenes* infection is 65 per cent. Approximately one-third of patients who have rheumatic fever (one-half of those with carditis) develop rheumatic heart disease, and severe disease results from recurrent attacks. The mitral valve is affected in nearly all patients, the aortic valve in 30 per cent and the triscuspid in 15 per cent (always accompanied by mitral valve disease).

Diagnosis

The diagnosis is based upon fulfilling certain clinical criteria and showing recent streptococcal infection. The modified Duckett Jones criteria are split into 'major' and 'minor'. Either two major, or one major and two minor, criteria are required.

- The five 'major' criteria are carditis, arthritis, chorea, erythema marginatum and subcutaneous nodules
- The five 'minor' criteria are fever, arthralgia, previous rheumatic fever or heart disease, raised acute phase reactants (ESR, C-reactive protein, leucocytosis) and prolonged PR interval
- Evidence for recent *S. pyogenes* infection needs to be provided by a

positive throat culture, increased ASO titre or other anti-streptococcal antibodies (e.g. anti-DNAse or anti-hyaluronidase), or a history of recent scarlet fever.

Treatment

- Bed rest if carditis present
- Aspirin or, in severe disease, corticosteroids
- Penicillin to eradicate pharyngeal *S. pyogenes*.

Prevention

- Secondary prophylaxis with oral penicillin V or monthly benzathine penicillin G. This is essential in preventing recurrences
- Primary prophylaxis by treating *S. pyogenes* tonsillopharyngitis with penicillin V
- There is no vaccine.

Prognosis

Mortality in the initial attack is very rare (<1 per cent) and results from severe carditis. One-third progress to develop rheumatic heart disease which has a significant chronic morbidity and mortality.

CHAPTER 9

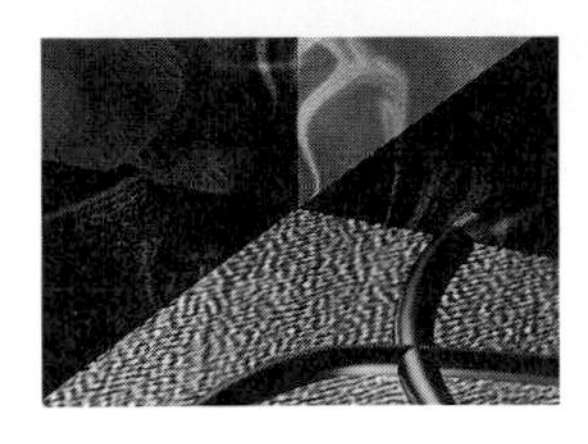

Infections of the Nervous System

CLINICAL SYNDROMES

Acute meningitis

Major presenting features: *rapidly developing headache, fever, meningism, photophobia.*

Three common types are recognised: pyogenic, aseptic and tuberculous. *Cryptococcus neoformans* is an important cause of meningitis in the immunocompromised.

Causes

Pyogenic

- Neonates: *Escherichia coli*, group B streptococci, *Proteus mirabilis*, *Pseudomonas*, *Listeria monocytogenes*
- Children: *Haemophilus influenzae*, *Neisseria meningitidis*, *Streptococcus pneumoniae*
- Adults: *N. meningitidis*, *S. pneumoniae*
- The elderly: *S. pneumoniae*, *Staphylococcus aureus*, Gram-negative enteric bacilli.

Aseptic

• Enteroviruses, mumps, herpes simplex, herpes zoster, adenoviruses, arboviruses. A similar cerebrospinal fluid (CSF) picture can be caused by early tuberculosis, syphilis, Lyme disease, encephalitis, brain abscess, sarcoidosis, systemic lupus erythematosus (SLE) and partially treated bacterial meningitis.

Tuberculous (see Plate I, facing p. 180)

• *Mycobacterium tuberculosis.*

Distinguishing features

	Pyogenic	*Aseptic*	*Tuberculous*
Onset	acute (<2 days)	acute (<2 days)	subacute (>2 days)
Clinical	toxic and ill, drowsy, purpuric rash (meningococci)	not toxic, fully conscious	not toxic, alertness may be depressed
CSF			
appearance	turbid or opalescent	clear	clear, may form cobweb on standing
predominant cells and number/μl	polymorphs, 500–2000	mononuclear, 5–1000	mononuclear, 50–400
protein (g/litre)	1–5	<1–5	1–3
Gram stain	usually positive	negative	negative

Complications

Pyogenic: deafness, blindness, cranial nerve palsies, hydrocephalus, subdural haematoma, cerebral abscess, Waterhouse–Friederichsen syndrome (meningococci), intellectual impairment

Viral: none

Tuberculous: hydrocephalus, hemiparesis, blindness, intellectual impairment.

Investigations

• Lumbar puncture (LP): essential if meningitis is suspected, unless there are signs of increased intracranial pressure (ICP)

• Computerised tomography (CT) scan: if ICP raised or focal neurological signs present, this should be performed before LP

• Gram stain of CSF smear

- CSF stain for acid-fast bacilli (if tuberculosis is suspected)
- Blood culture
- Demonstration of antigen in CSF/blood (by countercurrent immunoelectrophoresis (CIE) or Latex agglutination) if Gram stain is negative in pyogenic meningitis
- Faecal, throat swab and CSF viral cultures and viral serology in all cases of aseptic meningitis.

Diagnosis

Pyogenic

- Typical CSF picture
- Identification of organism from CSF Gram stain, CSF culture, antigen detection in CSF/blood.

Aseptic

- Enterovirus: isolation from faeces, CSF, throat swab
- Mumps: isolation from CSF, urine, serology
- Arbovirus: serology.

Tuberculous

- Acid-fast bacilli in CSF smear (in a minority)
- CSF polymerase chain reaction (PCR)
- CSF culture.

Treatment

Urgent empirical antibiotic therapy must begin in all cases of pyogenic meningitis, and treatment modified subsequently by the CSF findings. Intravenous (IV) cefotaxime or ceftriaxone will cover the three most likely organisms, i.e. meningococci, *H. influenzae* and pneumococci. An injection of penicillin should be given in suspected cases *before* the journey to hospital. In established cases the drugs of choice are:

- smear-negative pyogenic, *H. influenzae*, *S. pneumoniae:* IV cefotaxime or ceftriaxone
- aseptic: nil
- tuberculous: isoniazid, rifampicin, pyrazinamide
- *Listeria:* IV ampicillin
- *Neisseria meningitidis:* IV penicillin
- neonatal: ceftazidime and ampicillin.

Encephalitis

Major presenting features: *fever, evidence of cerebral dysfunction.*

Causes

Acute

- HERPES SIMPLEX, ENTEROVIRUS, ARBOVIRUS, rabies, Epstein–Barr virus (EBV), cytomegalovirus (CMV), human immunodeficiency virus (HIV), influenza, mumps, herpes zoster, *MYCOPLASMA*
- Post-exanthemata (measles, rubella, varicella).

Subacute or chronic

- HIV, subacute sclerosing panencephalitis (SSPE), progressive multifocal leucoencephalopathy (PML), CMV.

Differential diagnoses

- Toxic confusion in systemic infection
- Cerebral malaria
- Intracranial abscesses
- Tuberculous meningitis
- Subarachnoid haemorrhage.

Investigations

- CSF examination: changes are similar to those found in aseptic meningitis
- CT and/or magnetic resonance (MR) scans, electroencephalography (EEG): helpful in differentiation from brain abscess
- Brain biopsy: in selected cases of suspected herpes simplex virus (HSV) encephalitis or PML
- Demonstration of intrathecal antibody production (consult laboratory)
- Virus isolation from faeces, throat swab or CSF; PCR of CSF for HSV and serology.

Treatment

- IV acyclovir if HSV is likely, otherwise nutritional and cardiorespiratory support
- Treatment of convulsion; dexamethasone to relieve raised ICP.

Complications

- Brain damage
- Epilepsy.

Brain abscess

Major presenting feature: *cerebral dysfunction of acute onset.*

Causes

- BACTERIOIDES FRAGILIS and ENTEROBACTERIACEAE: often present together
- STAPHYLOCOCCUS AUREUS
- STREPTOCOCCUS MILLERI
- *Streptococcus pneumoniae*
- Anaerobic streptococci
- *Toxoplasma gondii* (in immunocompromised).

Diagnosis

- Clinical: apathy, inattention, signs indicative of focal cerebral or cerebellar defects
- CSF: clear fluid (LP should not be done if brain abscess is suspected), 0–50/μl mainly polymorph cells, normal biochemistry
- CT/MR scans
- Evidence of a source of infection, e.g. otitis, sinusitis, pneumonia, endocarditis.

Treatment

- Empiric antibiotic therapy with metronidazole, ampicillin and cefotaxime
- Surgery.

Polio-like illness/myelitis/spinal subdural abscess/ polyneuritis

These conditions may present in a similar manner. Major presenting feature: *rapidly developing weakness of limbs and trunk.*

Causes and distinguishing features

	Aetiology	*Clinical*	*CSF*
Polio-like illness	POLIO, coxsackie and echo viruses	Asymmetrical muscle weakness of purely motor type Sensory changes absent	Cells: 5–500 (usually mononuclear) Glucose: normal Protein: normal or slightly raised Manometry: normal
Myelitis	IDIOPATHIC, HSV, Zoster, CMV, EBV,	UMN weakness of lower limbs	Cells: 5–200 (usually mononuclear)

	Aetiology	*Clinical*	*CSF*
	post-exanthemata, HIV, HTLV I, Lyme borreliosis, schistosomiasis	Sensory level in the trunk	Protein: slightly raised Glucose: normal Manometry: normal
Spinal subdural abscess	*S. AUREUS*, *M. TUBERCULOSIS*, *Salmonella*	Back pain Spinal tenderness UMN weakness below a level in trunk	CSF: xanthochromic Cells: 20–200 (polymorphonuclear) Protein: markedly raised Glucose: normal Manometry: spinal block (Queckenstedt's phenomenon)
Guillain–Barré syndrome	IDIOPATHIC, BEV, CMV, post-exanthemata	Progressive, ascending, symmetrical weakness of limbs and trunk of LMN type Sensory symptoms are common Areflexia	Cells: <5 Protein: very high Glucose: normal Manometry: normal

Investigations

- CSF examination (not if spinal abscess is suspected, in which case consult neurosurgeons)
- Spinal X-ray
- MR scan
- Myelography
- Nerve conduction study
- Virus isolation from faeces, throat swab, CSF, serology
- In selected cases: serology for Lyme borreliosis, search for schistosomiasis.

Treatment

- Polio-like illness: bed rest, respiratory support, physiotherapy; orthopaedic appliances and surgery may be needed later

- Myelitis: bed rest, rehabilitation
- Spinal abscess: surgery
- Guillain–Barré syndrome: consider plasmapheresis or IV immunoglobulin.

SPECIFIC INFECTIONS

Meningococcal infection

Epidemiology

Meningococcal meningitis is the most common cause of bacterial meningitis worldwide. The causative organism is *N. meningitidis*, a Gram-negative intracellular diplococcus. Recognisable pathogenic groups are A, B, C, D, X, Y, Z and W 135. Group A organisms cause epidemics in West Africa, Sudan, Ethiopia, East Africa and the Middle East as well as in Nepal and India. In western countries, group B is the predominant organism, followed by group C.

Children and young adults are commonly affected. The organism is carried asymptomatically in the nasopharynx of 2–25 per cent of people and is spread by respiratory droplets from person to person. This is encouraged in overcrowded communities.

Pathogenesis

Clinical disease results from systemic bloodstream invasion with or without meningeal involvement. Waterhouse–Friderichsen syndrome is probably caused by septicaemic shock and disseminated intravascular coagulation with bleeding into and dysfunction of many organs of the body, including the adrenals. Group-related immunity follows an infection, even if subclinical.

Clinical features

The incubation period (IP) is 1–3 days. The disease starts abruptly with fever, headache, irritability and restlessness, rapidly progressing to signs of meningitis. The illness may also present just as septicaemia with rapidly developing toxicity, drowsiness and shock. A petechial or purpuric rash is present in two-thirds of cases.

Complications

- Waterhouse–Friderichsen syndrome: fulminant septicaemia with evidence of adrenocortical failure

- Hydrocephalus, brain damage, subdural haematoma, brain abscess, deafness
- Chronic meningococcal septicaemia presenting with bouts of pyrexia and recurrent rashes over many weeks (rare)
- Arthritis (septic or reactive), cutaneous vasculitis, pericarditis.

Diagnosis

- In patients with signs of disturbed cerebral function, LP should be withheld until a CT scan has excluded cerebral oedema
- CSF changes are usually characteristic of pyogenic meningitis (see p. 114), and Gram stain shows Gram-negative diplococci, often intracellular. In the septicaemic form, the CSF cells may be normal or minimally raised. Gram stain may be negative even in the absence of prior antibiotic therapy, so CSF and blood cultures are essential. Antigen detection in CSF and blood (Latex agglutination, CIE) are helpful for rapid diagnosis in Gram stain negative cases.

Treatment

- Petechial rash present: IV penicillin, 10 days
- Petechial rash absent: IV cefotaxime until diagnosis confirmed, then IV penicillin, total 10 days; IV chloramphenicol if allergy to penicillin
- Septicaemia: ill patients should have haemodynamic monitoring, with cardiorespiratory support if necessary.

Prognosis

The overall mortality is 5–10 per cent of cases mostly from rapidly progressive fatal septicaemia.

Prevention

- All cases must be notified and chemoprophylaxis given to all close contacts (household contacts, contacts in day-care centres): either rifampicin 600 mg twice daily for 2 days (children 10 mg/kg) or ciprofloxacin (500 mg single dose in adults) or single injection of ceftriaxone
- Nasopharyngeal swabbing is not necessary for prevention purposes
- Chemoprophylaxis will not prevent illness in those already incubating it (co-primary cases) and close surveillance of intimate contacts are important
- Vaccines containing polysaccharide antigens of A and C are available and effective. These are used in times of epidemics to control spread and for vulnerable groups like military recruits and travellers to endemic areas of subsaharan Africa and the Indian subcontinent. Conjugated group C vaccines are being tested for children under 2 years old.

Haemophilus meningitis

Epidemiology

• *Haemophilus influenzae* organisms are common in the respiratory tract of both children and adults but, of the six antigenically distinct capsular types (a, b, c, d, e and f), only type b produces meningitis

• Possible inflammation of the upper respiratory passages occurs initially, followed by invasion of the bloodstream, with involvement of meninges hours or days later

• Transmission is via respiratory route. Cases mainly occur in children from 3 months to 5 years of age with a peak incidence around 1 year of age. Older children and adults are immune because of previous, often subclinical, infections

• The incidence of invasive *H. influenzae* disease has declined dramatically in countries which have introduced conjugated *H. influenzae* type b (Hib) vaccine.

Clinical features

The disease often presents less acutely than either meningococcal or pneumococcal meningitis. A period of fever and malaise often precedes the development of signs of meningeal irritation. Untreated, the patient will become drowsy and eventually comatose, sometimes with convulsions and cranial nerve paralysis.

Diagnosis

Lumbar puncture usually reveals a CSF typical of pyogenic meningitis (see p. 114). The Gram stain shows Gram-negative pleomorphic coccobacilli and CSF and/or blood culture are usually positive. The bacterial antigen can usually be detected by CIE in Gram-negative cases. There is polymorphonuclear leucocytosis in peripheral blood.

Complications

Cranial nerve palsies, hydrocephalus, deafness, particularly in cases of delayed treatment and subdural effusion.

Treatment

Ampicillin- and chloramphenicol-resistant strains, are not uncommon and a third-generation cephalosporin such as cefotaxime, cefuroxime or ceftriaxone is now the drug of choice. Dexamethasone as an adjunctive therapy helps to prevent deafness.

Prognosis

Effective early treatment reduces the mortality almost to 1 per cent of cases. Complete recovery is usual but long-term neurological sequelae may occur if treatment is delayed.

Other diseases produced by *H. influenzae* type b

These occur in young children only.

- Acute epiglottitis
- Septic arthritis/osteomyelitis
- Cellulitis affecting the face, head or neck. The onset is abrupt and the affected skin often assumes a bluish-purple appearance, which may be mistaken for a bruise
- Pneumonia, empyema and pericarditis.

Prevention

- Consider prophylaxis (rifampicin 20 mg/kg daily for 4 days) if there are other children in the household below the age of 4 years. Similar action should be taken in day-care centres if two or more cases occur
- A protein–polysaccharide conjugate vaccine gives effective protection against *H. influenzae* type b diseases and is recommended for routine use in all children concurrently with the diphtheria/tetanus/pertussis vaccination, and for any unvaccinated young child contacts of a case.

Pneumococcal meningitis

Epidemiology

Although less common than either meningococcal or *Haemophilus* meningitis, pneumococcal meningitis is a more severe disease with a high mortality rate. The source of infection may be haematogenous from respiratory tract or an unknown source, or direct (from a chronically infected middle ear or via a congenital defect or skull fracture). The disease is not infectious. Although it is not uncommon in young infants, it is the most common type of bacterial meningitis in the over 50 age group. Asplenic and hypogammaglobulinaemic patients are high-risk groups.

Clinical features

The illness starts acutely with fever, headache, neck stiffness, photophobia and vomiting, progressing rapidly to drowsiness and coma and sometimes convulsions. In patients with recurrent pneumococcal meningitis, there may be a history suggestive of CSF rhinorrhoea or otorrhoea.

Complications and prognosis

Cranial nerve damage, ventriculitis and hydrocephalus (these are more common in pneumococcal meningitis than in other types of bacterial meningitis), subdural haematoma and cerebral abscess. Mortality is high even with treatment and may approach 20 per cent in the older age group.

Diagnosis

- CSF shows typical changes of pyogenic meningitis (see p. 114). Gram stain of CSF deposit usually shows Gram-positive diplococci. CIE for pneumococcal antigen is usually positive in CSF/serum and sometimes in urine. CSF and blood should be cultured for isolation and determination of antibiotic susceptibility
- There is usually a polymorphonuclear leucocytosis in peripheral blood.

Treatment and prevention

- High-dose IV penicillin or a third-generation cephalosporin such as cefotaxime are drugs of choice. In penicillin-allergic patients, chloramphenicol should be used
- Patients with an increased risk of severe pneumococcal disease (those with asplenia, diabetes, immunosuppression or heart, lung or kidney diseases) should receive pneumococcal vaccine.

Neonatal meningitis

Epidemiology

The two main causes of neonatal meningitis are *E. coli* and group B streptococci, and less commonly, *L. monocytogenes*, *Proteus mirabilis*, *S. aureus* and *P. aeruginosa*. Up to one-quarter of women of child-bearing age carry group B streptococci in the lower genital tract or rectum. Low birth weight pre-term babies and those with congenital anomalies such as spina bifida or receiving intensive care management are particularly vulnerable. Cross-infection may occur in special care baby units involving coliforms, staphylcocci and group B streptococci.

Clinical features

Presentation is usually nonspecific, lacking the classical signs of meningitis. Pallor, irritability, vomiting and failure to feed and thrive may be the only signs; jaundice and respiratory distress may be present. A bulging fontanelle is present in a minority of cases.

Septicaemia is often present. The disease is rapidly progressive and often fatal in the early-onset type of illness due to group B streptococci (occurring within a few days of birth), whereas the illness occurring in the 2nd and 3rd weeks of life presents more insidiously.

Diagnosis

CSF examination is of less value in neonates than in older children, as cell counts up to 30/μl and protein levels up to 1.5 g/litre are not uncommon in high-risk infants without meningitis. Gram stain of CSF deposit and culture of CSF and blood are more helpful.

Treatment

Empiric antibiotic therapy should begin urgently on suspicion of neonatal meningitis, once blood and CSF samples have been collected. The combination of ceftazidime and ampicillin is widely used.

Prognosis

Overall mortality is high in neonatal meningitis (up to 50 per cent of cases in prematurely born infants) and there is a high incidence of neurological sequelae in up to one-third of the survivors.

Listeriosis and *Listeria* meningitis

Epidemiology

Listeria monocytogenes is a Gram-positive bacillus present in the environment (water and earth) which commonly infects humans and animals, leading to asymptomatic faecal excretion. Vaginal carriage may occur in women. Sources of human infection are unpasteurised milk, cheese, paté, contaminated vegetables, or other food or infective material (e.g. aborted fetus). Neonatal listeriosis can be acquired from an infected maternal birth canal.

All age groups are susceptible but newborns, pregnant women, elderly and immunocompromised or otherwise debilitated persons have a higher incidence. The IP varies from 3 to 7 days.

Clinical features

Infection is often asymptomatic in the normal host or may only produce a short-lasting influenza-like illness. In pregnant women, transient bacteraemia of mild illness can still cause fetal infection leading to abortion or a stillborn or badly damaged baby. Occasionally septicaemic (mainly in children) or meningitic (mainly in adults) illness may develop.

In meningitis, onset is typical of pyogenic meningitis but, in the immunocompromised, the disease may present subacutely. The CSF usually has the typical picture of pyogenic meningitis but sometimes the cellular response may be lymphocytic and may mimic tuberculous meningitis because of a low glucose level. Neonates may have a septicaemic illness at birth, or meningitis later in the neonatal period.

Diagnosis

This is by isolation of the organism from CSF, blood, meconium and gastric washings. Microscopy shows short Gram-negative rods which may occasionally be confused with diphtheroids.

Treatment

The drug of choice is ampicillin, which may be combined with an aminoglycoside. Chloramphenicol is a suitable alternative for penicillin-allergic patients. Mortality is high (30 per cent or more).

Prevention

Pregnant women and the immunocompromised should avoid animal contact on farms and unpasteurised milk or milk products, or inadequately cooked meats. High-risk foods, e.g. soft cheese and paté, should be stored below 4°C. Salads should not be eaten beyond their 'use by date'. Dairy products should be monitored for *Listeria* if pasteurisation is not possible.

Tuberculous meningitis

This may be an early or late complication of primary tuberculosis, or may accompany miliary and chronic pulmonary tuberculosis.

Clinical features

- The onset is insidious with mild recurrent headache, slight malaise and fever
- After a week or longer there is more severe headache and pyrexia, mental and personality changes or progressive drowsiness. Neck and spine rigidity appears and cranial lesions, particularly of the 6th and 3rd nerves occur
- During the 3rd or 4th week, coma ensues and hemiplegia and diplegia often develop. Hyponatraemia is common.

Diagnosis

The salient clinical and CSF features which differentiate tuberculous meningitis from bacterial and viral meningitis are shown on p. 114. Acid-fast bacilli can be seen in CSF in a minority of cases. Culture is normally positive but takes time. Rapid confirmation can be obtained by DNA amplification (PCR).

Tuberculous meningitis needs differentiation from:

- viral meningitis
- cryptococcal and *Listeria* meningitis (some cases), because of low glucose and a predominantly lymphocytic CSF
- partially treated bacterial meningitis if low CSF glucose persists, when cells have reduced in number and have become predominantly lymphocytic. Such cases are rare
- other causes of subacute meningitis, i.e. sarcoidosis, neurosyphilis, leukaemia.

Treatment

As for pulmonary tuberculosis (see p. 99), except that the duration should be for 9 months.

Viral meningitis (aseptic meningitis, lymphocytic meningitis)

This is a benign self-limiting disease which may be caused by a variety of viruses (see p. 113), but in the UK most cases are due to enteroviruses (coxsackie and echo viruses).

Epidemiology

Occurrence is worldwide with sporadic cases as well as epidemic outbreaks. Enteroviral meningitis mainly occurs during the summer months. Mumps meningitis was common in the UK before the introduction of routine mumps immunisation. Poliovirus and arboviruses are important causes of meningitis in different parts of the world but are rare in the UK. Acute seroconversion illness due to HIV infection may present as meningitis. Lymphocytic choriomeningitis (LCM) is now almost extinct in the UK, but occurs in other parts of the world including Europe and the Americas. The LCM virus is an arenavirus; human infection occurs through contact with infected rodent urine.

Children and young adults are mainly affected.

Clinical features

The onset is acute with fever, headache, vomiting and signs of neck and spinal rigidity. A faint pink, maculo-papular rash (often rubella-like) may be present with some enteroviral infections. Presence of parotitis in usual in mumps meningitis but may be absent. Headache in viral meningitis may be intense and may persist for a week even after subsidence of fever and neck stiffness. In general, patients are much less ill-looking and non-toxic in comparison with those with bacterial meningitis. Consciousness is unimpaired unless there is associated encephalitis.

Diagnosis

• CSF shows typical changes of aseptic meningitis (see p. 114). Mumps infection is diagnosed by isolation of the virus from the throat swab, urine or CSF, or by demonstrating rising antibody titre in paired sera; enterovirus infection is diagnosed by the isolation of the virus from faeces, CSF or throat swab

• Differentiation is from early tuberculous meningitis, rare cases of bacterial meningitis where the CSF picture has become modified from pre-admission antibiotic therapy, brain abscesses and other forms of parameningeal sepsis, fungal and leptospiral infections.

Treatment and prognosis

Recovery is always complete without any treatment.

Poliomyelitis

Epidemiology and pathogenesis

• Poliomyelitis is caused by poliovirus, which is an enterovirus. It exists in three antigenically distinct types – 1, 2 and 3. Type 1 is the most virulent. Transmission is faecal–oral and, rarely, respiratory. In most industrialised countries the disease has become rare because of widespread use of vaccine, but the virus is still prevalent in many developing countries, where infection occurs early in life and most adults are immune

• The IP is 7–14 days. After penetrating the intestinal mucosa, the virus spreads to the regional lymph nodes. In some patients further multiplication at these sites leads to viraemia which coincides with the onset of fever. Virus may then localise in the meninges, causing meningitis, and less frequently in the motor nuclei of the spinal cord or brain causing poliomyelitis or polioencephalitis.

Clinical features

The outcome of infection may be asymptomatic (most common), or a short nonspecific febrile illness or meningitis with or without paralytic illness. Paralysis occurs only in a small minority of those infected.

Meningitis and poliomyelitis

- These present acutely with fever, headache, vomiting and signs of meningeal irritation. In many cases the disease does not progress further and settles in a few days, but in others, muscular pain followed by flaccid paralysis develops
- The paralysis is asymmetric in distribution and may continue to spread for several days until the patient's temperature returns to normal. Sensory changes do not occur. The legs, and to a lesser extent the arms, are most affected
- When the brainstem nuclei are involved, bulbar paralysis with weakness of swallowing and coughing develops. Rarely, there may be encephalitis from involvement of the motor cortex
- Virus excretes in the faeces for several weeks after the cessation of fever.

Complications

- Respiratory failure occurs from paralysis of the muscles of respiration or brainstem involvement
- Recovery of muscle power is rarely complete and, in severe cases, extensive paralysis persists, leading to atrophy and deformity.

Diagnosis

- The CSF shows changes typical of viral meningitis (see p. 114). Acute polyneuritis and myelitis are other causes of rapidly developing paralysis
- Confirmation of diagnosis is obtained by isolation of the virus from faeces and also from the nasopharynx but rarely from CSF, and by serology showing a four-fold rise in antibody titre. Rarely, coxsackie and echo viruses may cause a polio-like illness.

Treatment

Complete bed rest, with control of fever and pain during the early stage; the main measures are ventilatory support if respiratory failure develops, and physiotherapy when spread of paralysis ceases. Later, orthopaedic aids and surgery may be needed.

Prognosis

Respiratory failure accounts for most deaths (2–10 per cent incidence).

Adults tend to suffer from severe disease with increased fatality. Late deaths occur in patients suffering from permanent respiratory insufficiency.

Prevention

- In hospital, patients should be nursed under enteric precautions, but isolation at home is of little value as dissemination has occurred already during the prodromal stage
- In developed countries, where the incidence of poliomyelitis is very low, the source of infection in every single case needs investigation by searching among contacts (unless infection has been contracted abroad)
- Immediate vaccination of the surrounding population with a trivalent live oral vaccine will limit the spread of the virus to uninfected individuals, although vaccination has no value for close contacts who are already infected
- Minor operations and prophylactic injections are postponed among contacts as these are believed to precipitate paralysis
- Polio immunisation: two types of polio vaccines are available – inactivated poliovirus (IPV) and live attenuated oral poliovirus (OPV) vaccines. Both are highly effective. Their use varies in different countries: OPV vaccine is used by some countries, IPV vaccine by some, and others use both. In the UK, three doses of trivalent OPV vaccine (contains all three virus types) are given concurrently with routine childhood diphtheria, tetanus and whooping cough vaccines. Additional doses are given at school entry and at the age of 15 years. Further doses are given at 10-yearly intervals to people travelling to endemic areas. For endemic countries, the World Health Organisation recommends an additional dose given at birth (because of the early occurrence of paralytic poliomyelitis in these areas)
- OPV vaccine colonises the gut, producing local as well as humoral immunity, and also spreads to susceptible contacts, protecting them. Rarely, OPV vaccine can cause paralytic illness (1 case per 2.6 million doses) either in the recipient or in their healthy contacts, particularly in adults. In the USA, IPV vaccine is often preferred for exposed, non-immunised adults. It is also the vaccine of choice when OPV vaccine is contraindicated, i.e. for individuals who are immunosuppressed or whose household contact is immunosuppressed
- Both OPV and IPV vaccines contain minute traces of penicillin and neomycin but, except in cases of extreme hypersensitivity, OPV vaccine can be used safely.

Other enteroviral infections

Enteroviruses belong to the family of Picornavirus and, apart from polio, other recognised human enteroviruses are coxsackie (24 group A and six group B serotypes), echo (34 serotypes), 'new' enteroviruses (types 68–71) and hepatitis A virus. A wide range of clinical syndromes are caused by these agents. They all infect the human gut and viraemia is common.

Epidemiology

Occurrence is worldwide and transmission is usually faecal–oral. The IP is usually 2–5 days (except for hepatitis A). Asymptomatic infections are very common.

Clinical features

The following clinical syndromes may occur, but there is considerable overlapping:

- aseptic meningitis
- Bornholm disease (epidemic myalgia, pleurodynia). Caused by group B coxsackie viruses, this condition is characterised by severe lower intercostal or upper abdominal muscular pains, associated with fever and malaise. The pain fluctuates in intensity and may be pleuritic. Symptoms usually subside within 1 week but may relapse
- simple febrile illness, or undifferentiated upper respiratory tract infections (coxsackie and echo viruses)
- herpangina: painful tiny greyish ulcers over the soft palate and fauces (coxsackie A)
- fever with maculo-papular rash (coxsackie and echo viruses)
- hand–foot–mouth disease, characterised by vesicles on hands and feet and ulcers in the mouth in young children (mostly coxsackie A16)
- pericarditis or myocarditis: occur rarely in adults and teenagers (coxsackie and echo viruses)
- neonatal infection: severe disseminated infection with myocarditis, hepatitis and encephalitis (mainly echo virus)
- flaccid paralysis: a lower motor neuron paralysis, similar to poliomyelitis (coxsackie and echo viruses)
- enteroviral haemorrhagic conjunctivitis: this is usually caused by enterovirus 70 and occasionally by a variant of coxsackie A24. Enterovirus 70 has caused large outbreaks of conjunctivitis in many areas of India, South-East Asia, Africa, Central and South America. The virus is transmitted by hands and contaminated towels. The symptoms appear within 2 days of exposure to an infected person. Subconjunctival haemorrhage is common. Recovery is usual within 10 days.

Diagnosis and treatment

This requires isolation of the virus from faeces, and less commonly from throat swab or CSF (or from conjunctiva if eyes are involved). Serology can be useful to confirm diagnosis for some syndromes. There is no effective treatment.

Encephalitis

Encephalitis is a diffuse inflammatory condition of the brain and most cases are generally assumed to be caused by a virus infection. Two pathologically distinct forms of acute encephalitis are recognised: primary and post-infective.

Primary encephalitis

- This is due to a direct virus attack on the brain, as caused by herpes simplex virus (HSV), enteroviruses, rabies, arboviruses, HIV, influenza, EBV and CMV. There is neuronal inflammation with perivascular infiltration of inflammatory cells. The grey matter is predominantly involved
- In the UK, primary encephalitis is rare and usually sporadic. HSV is responsible for a proportion of these cases and enteroviruses account for a few. Rarer causes include EBV, *Mycoplasma*, adenovirus and influenza virus. HIV can cause both acute encephalitis during the initial infection and chronic encephalopathy in late stages. CMV has accounted for a few cases of encephalitis in individuals with severe HIV-induced immunodepression. Often, no aetiological agent can be identified
- Outbreaks of encephalitis due to various arboviruses occur in many parts of the world, particularly Japan, Russia, South America, Australia and the USA.

Post-infective meningo-encephalitis

This is a rare complication of measles, chickenpox and rubella. A similar illness occurs after the use of nerve tissue-containing rabies vaccine and, previously, following smallpox vaccination. The illness usually occurs several days after the onset of the viral infection. The condition is believed to be due to an immunological reaction triggered by the virus and is characterised pathologically by perivenous demyelination, with the white matter being predominantly affected. There may be associated myelitis.

Clinical features and diagnosis

- The onset of encephalitis is acute, with fever, headache, vomiting, irritability and photophobia

- There may be fluctuations in the level of consciousness and changes in the personality, with confused and abnormal behaviour. Later, the patient becomes progressively drowsy and even comatose
- Involuntary movements, twitching or frank convulsion may appear
- In HSV encephalitis, focal signs such as hemiplegia or aphasia are common, but are unusual in other types
- The CSF usually shows changes typical of aseptic meningitis (see p. 114), with normal glucose and slightly raised protein, but rarely may be normal
- The EEG generally shows a diffused bilateral slow wave pattern without any focal features, except in HSV encephalitis
- The clinical picture, and CSF and EEG changes, are usually sufficient to distinguish encephalitis from the many toxic and metabolic conditions, but CT scan should be performed to exclude intracranial abscess. Tuberculous meningitis must be excluded. Biopsy may be necessary in difficult cases
- Virus isolation should be attempted from throat swab, faeces, urine and CSF in all cases of primary encephalitis, and paired samples of sera should be collected for serological studies against a wide range of possible viral agents
- Demonstration of HSV by PCR technique and CSF–serum antibody ratio are helpful in diagnosing HSV encephalitis.

Treatment

- The treatment of encephalitis is generally entirely supportive, with maintenance of vital functions and measures to reduce raised ICP (dexamethasone). In suspected or proven HSV encephalitis, IV acyclovir reduces mortality and morbidity
- Corticosteroids are not helpful in post-infectious encephalitis.

Prognosis

- Mortality varies from 0 to 30 per cent depending on the type
- Neuropsychiatric sequelae are not uncommon.

Subacute and chronic encephalitis

Subacute sclerosing panencephalitis

- This is a very rare form of encephalitis in children and adolescents, characterised by mental deterioration and myoclonic fits of insidious onset, gradually progressing to generalised convulsion, coma and probable death
- It is a complication of measles starting months, or more usually several years, after the acute infection (1/1000,000 cases). There is persistence

of measles virus in the brain, probably due to some abnormal immunological response. It is now very rare in countries with routine childhood measles vaccination programmes
• The diagnosis is by the typical clinical features, characteristic EEG changes (wave and spike forms) and high measles antibody titre in CSF. There is no treatment.

Progressive multifocal leucoencephalopathy

• This is due to a papovavirus infection, typically the JC virus. There is demyelination with abnormal astrocytes and oligodendrocytes, affecting predominantly the parieto-occipital white matter. PML occurs only in patients with depressed cell-mediated immunity and often affects patients with acquired immunodeficiency syndrome (AIDS)
• Clinically, there is progressive limb weakness followed by disturbances of cognitive, visual and speech functions, lack of coordination of limbs, and headache, ending fatally in a few months
• MR scans, showing characteristic white matter lesions, help diagnosis. CSF is normal but PCR is positive for JC virus
• No effective therapy is available.

CMV and HIV

• These are other causes of chronic encephalitis seen in immunosuppressed persons.

Acute polyneuritis (Guillain–Barré syndrome)

This is a rare, sporadic disease of adults and older children, characterised by rapidly progressive paralysis of limbs and trunk muscles.

Epidemiology and pathogenesis

An infective trigger is often suspected but infrequently proven. EBV and CMV account for a few cases. An outbreak occurred in the USA associated with influenza vaccine in 1976. It may occur as a rare complication of chickenpox or measles and following the use of nerve tissue rabies vaccine. An infection-triggered immune-mediated damage to myelin sheaths has been postulated. A preceding upper respiratory infection or sore throat occur in about one-quarter of patients. Occurrence is worldwide.

Clinical features and diagnosis

• The illness is characterised by a slowly progressive symmetrical lower motor neuron type paralysis of the limbs and trunk, with symmetrical distal sensory loss. There is areflexia. Paralysis of respiratory and facial muscles and bulbar paralysis develop in severe cases

• The differentiation is from diphtheritic polyneuritis, poliomyelitis and acute myelitis (see p. 117)

• The CSF shows greatly raised protein without any pleocytosis or abnormal glucose.

Treatment and prognosis

• Complete recovery is usual, though this may take a long time. Deaths are from complications of intensive therapy. A few patients have spinal cord involvement (myeloradiculitis) and suffer from permanent residual weakness

• Plasmapheresis and IV immunoglobulin, used early, may be beneficial

• Respiratory failure requires management in an intensive care unit with assisted ventilation and establishment of airway

• Glucocorticosteroids are not helpful.

Reye's syndrome

This is an encephalopathic illness in which a previously healthy child (usually below 2 years old) develops convulsions and fever, then lapses quickly into status epilepticus and coma. The liver enlarges rapidly and may fail. Hypoglycaemia occurs. Fatality is common.

Though frequently confused with encephalitis, CSF is normal (except for a low sugar) and at autopsy the brain is only swollen, with no histological evidence of encephalitis. The liver and sometimes the heart, kidney and pancreas, show fatty degeneration.

Defective fatty acid metabolism has been postulated and an association with various virus infections has been noted. Aspirin has been linked as a co-factor and, since its withdrawal from paediatric use, the incidence of Reye's syndrome has declined considerably in Britain.

Botulism

This is a paralytic illness caused by the neurotoxin of *Clostridium botulinum*.

Epidemiology and pathogenesis

• The organism is an anaerobic, spore-forming bacterium which is widespread in soil. The spores are heat-resistant and can survive up to 2 hours at 100°C but are killed at 120°C. The seven types of *C. botulinum* all produce neurotoxin, which paralyses muscles

• Most cases of human botulism are due to ingestion of toxin via food in which contaminating spores have been allowed to germinate in an anaerobic environment, thus producing the toxin, e.g. consumption of preserved or canned foods such as meat, fish, fruit and vegetables which have been inadequately heat treated. Rarely, botulism may be caused by contamination of wounds by *C. botulinum* with subsequent production of exotoxin and, in infants, by colonisation of the intestine leading to production and absorption of toxin

• Botulism is very rare in Britain. The last outbreak occurred in 1989 involving 27 cases, and was due to consumption of commercially produced hazelnut yoghurt.

Clinical features

• In food-poisoning botulism, after an IP of 12–36 hours, there is an onset of vomiting, tiredness, thirst and bulbar and ocular muscle paralysis. Within 24 hours, flaccid paralysis of the limb and trunk muscles develops, leading to death from respiratory failure in half of the cases. Sensation remains intact. The temperature is normal throughout. Recovery is gradual over a period of weeks or months, but complete

• In infant botulism, most cases have a mild and transient illness with constipation and lethargy as prominent features. In a few infants, illness is more severe with difficulty in feeding, paralytic squints, flaccid limbs and trunk weakness.

Diagnosis

• This is primarily clinical, at least in the initial stage. Differential diagnoses includes GBS, bulbar myasthenia of rapid onset, diphtheria with palatal and extraocular palsies, organophosphorous poisoning, cerebrovascular accidents involving the basilar artery, and paralytic shellfish poisoning

• Confirmation is by demonstrating toxin in the patient's blood (or faeces) by inoculating into mice, which will show signs of botulism.

Treatment and prevention

Botulinum anti-toxin is of doubtful value and treatment is generally supportive. Modern and efficient commercial standards of canning prevent botulism; most of the outbreaks worldwide are related to inadequate home preservation of food.

Tetanus

This is caused by *C. tetani*, which is a Gram-positive, anaerobic, spore-forming bacillus present in the bowels of many herbivorous animals and also of humans.

Epidemiology

• Spores of tetanus are widely distributed in soil, and in the absence of active immunisation a person of any age can develop tetanus through wounds contaminated with soil

• Because of routine childhood immunisation, the incidence of tetanus is very low in Britain (less than 40 cases/year). Elderly women are mostly affected (born before the introduction of childhood immunisation and not part of military service, where adults received vaccine)

• In developing countries tetanus remains an important cause of death. Neonatal tetanus is a special problem in some countries where mud or animal dung are used to treat the umbilical stump.

Pathogenesis

• In wounds contaminated with spore-bearing soil, anaerobic conditions created by the presence of foreign bodies and devitalised tissues encourage active vegetative growth of *C. tetani*, leading to toxin production (tetanoplasmin)

• The toxin travels proximally along the nerves to reach the nervous system. It produces tetanus by two mechanisms: by blocking acetylcholine release at the myoneural junction and by countering the inhibitory influences on muscle reflex arcs

• The resultant increase in lower motor neuron activity leads to the muscle rigidity and spasms which characterise tetanus. Once 'fixed' in the spinal cord, the toxin can no longer be neutralised by anti-toxin

• Removal of inhibitory influences on the autonomic nervous system causes increased autonomic activity, e.g. tachycardia, sweating and hypertension.

Clinical features

• The IP is normally 5–15 days, but may be longer. Short incubation periods are associated with severe disease

• Rigidity stage: trismus (painfully stiff jaw muscles) is often the first symptom, with difficulty in opening the mouth (lockjaw). Dysphagia may develop. Fever, if present, is mild. In 24 hours, stiffness spreads to neck, back, chest and abdominal wall muscles. Arms and legs are only slightly affected

• Spasmodic stage: generally within 1–2 days, intermittent, spasmodic, painful contractions of the affected muscles develop – often accompanied by pallor and sweating. The interval between the onset of rigidity and onset of spasm is known as the 'period of onset' – shorter periods are associated with severe tetanus

• The spasms cause grimacing of the face (risus sardonicus, see Plate 2, facing p. 180) and arching of the neck and back (opisthotonos). Spasms of laryngeal and respiratory muscles cause respiratory failure. The spasms occur spontaneously or may be triggered by noise, coughing and movements

• In severe cases signs of sympathetic overactivity appear: profuse sweating, fever, hypertension/hypotension, tachycardia, cardiac arrhythmia

• In surviving patients, the spasms gradually subside after 2–3 weeks and muscle rigidity disappears after another 1–2 weeks

• In mild cases, there is often rigidity alone, and this may rarely be localised only to the site of injury.

Diagnosis

This is usually clinical, although *C. tetani* may occasionally be found in the wound. Tetanus is extremely rare in an adequately immunised person. Presence of anti-toxin levels of 0.01 IU/ml in the admission sample in a patient with tetanus-like spasms excludes the diagnosis.

Differential diagnoses

• Metoclopromide and prochlorperazine or similar drugs may cause acute dystonia involving head and neck muscles. Rigidity is absent and oculogyric spasms are characteristic. The condition responds dramatically to 2 mg IV benztropine

• Trismus may be mimicked by dental abscess, mumps, temporomandibular joint problems

• Rabies, strychnine poisoning and spinal myoclonus.

Treatment

Ideally, a patient with muscle spasms should be treated in an intensive care unit. The important steps in management are:

• administration of human tetanus immunoglobulin 20,000 IU intravenously, followed by debridement of the wound

• benzylpenicillin IV or IM for 10 days to eradicate the existing foci of infection and stop further toxin production

• sedation of the patient: diazepam is used widely and may control mild spasms; patients should be nursed in a quiet environment to avoid triggering spasms

• all patients with generalised spasms, even if not severe, should have a tracheostomy to guard against sudden life-threatening laryngospasm

- if sedation is not controlling spasms, the patient should be paralysed with muscle-relaxing drugs like pancuronium and be ventilated
- attention to fluid and electrolyte balance and nutrition is important
- sympathetic overactivity may be helped by labetalol, which possesses both α- and β-blocking properties, but hypertensive crisis may require other drugs like diazoxide
- finally, the patients who recover from tetanus should be given active immunisation as the disease does not produce immunity.

Prevention

Active immunisation with tetanus toxoid gives protection for at least 10 years, probably much longer.

- The most important step is universal childhood immunisation with a three-dose primary course followed by boosters at school entry and school leaving. In developing countries, immunisation of women at antenatal clinics prevents neonatal tetanus
- All wounds should be thoroughly cleaned, with removal of foreign bodies and dead tissue. Penicillin or erythromycin prophylaxis for contaminated or infected wounds may reduce the likelihood of tetanus
- Patients with injuries should be considered for active and passive immunisation as detailed in Table 9.1.

Prognosis

Modern management reduces the mortality of severe tetanus from 60 to

TETANUS IMMUNISATION IN INJURY

	Type of wound	
Tetanus vaccination history	Clean, minor wounds <6 hours old	All other wounds
Uncertain, none, incomplete course	Give tetanus toxoid;* consider antibiotics	Give tetanus toxoid* and immunoglobulin†
Three or more doses:		
last dose >10 years ago	Give tetanus toxoid booster	Give tetanus toxoid booster
last dose <10 years ago	Nil	Nil

*Complete basic course of tetanus immunisation
†Human anti-tetanus immunoglobulin

Table 9.1. Active and passive immunisation against tetanus in injured patients

10–20 per cent. Most deaths are due to sympathetic hyperactivity. Mild or localised tetanus has no mortality.

Slow viral infections of the CNS (spongiform encephalopathies)

There is a group of slowly progressive, degenerative, non-inflammatory diseases of the brain affecting animals and humans which are thought to be caused by filtrable transmissible agents, known as 'prions'. They are often called spongiform encephalopathies.

Examples affecting animals

- Scrapie (sheep)
- Bovine spongiform encephalopathy (cattle)
- Transmissible encephalopathy of mink.

Forms recognised in humans

- Creutzfeldt–Jakob disease (CJD). This is characterised by progressive confusion, dementia and ataxia, usually in people between 40 and 70 years of age. Later, spasticity, wasting and sometimes myoclonus develop, ending in coma and death within a year from onset. CSF is normal. EEG commonly shows diffuse periodic high-voltage discharges. Histology of the brain shows characteristic changes and the diagnosis is confirmed if the disease can be transmitted to animals using brain biopsy materials. Mode of transmission is unknown. There are reports of transmission via the use of human pituitary and other tissues
- Kuru. This has occurred exclusively among the cannibalistic tribal people of New Guinea. A transmissible agent similar to CJD is involved. The disease is characterised by progressively worsening cerebellar symptoms and wasting, ending in death. It has become much less common due to changed ritualistic practices.

There is no evidence that the agents of animal spongiform encephalopathies can cause disease in humans.

CHAPTER 10

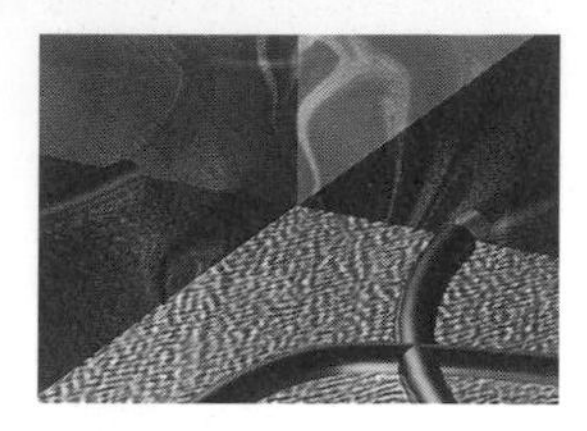

Skin and Subcutaneous Tissue Infections

CLINICAL SYNDROMES

In many infections the dominant clinical features are confined to the skin, with or without deeper soft tissue involvement. These conditions can be conveniently grouped under two broad headings:

- infections associated with a widespread rash which may be maculo-papular, erythematous, purpuric/haemorrhagic or papulo-vesicular (see Table 10.1 for definitions)
- infections associated with a localised involvement of the skin with or without deeper tissue involvement.

TYPES OF RASH

Rash	Size	Description
Macules	<5 mm diameter	Spots of abnormally coloured skin which fade on pressure
Papules	Pea size or smaller	Solid projections above the surface of the skin. Larger similar projections are called *nodules*
Vesicles	Pea size or smaller	Projections above the surface of the skin containing clear fluid. Larger similar projections are called *ballae* or *blisters*
Pustules	Small	Elevations of skin containing pus
Purparae	<1 mm diameter	Haemorrhages in the skin. Larger similar haemorrhages are called *ecchymoses*

Table 10.1. Types of rash

Infectious conditions associated with widespread rash

Condition	*Discriminating features*
Maculo-papular rash	
Measles	Prominent 3–4 days febrile prodrome with coryza, conjunctivitis and Koplik's spots. Rash slowly evolves down the body, becoming confluent
Rubella	Non-existent or very short-lasting prodrome. Rapidly evolving discrete rash. Enlarged post-cervical glands

Condition	*Discriminating features*
Erythema infectiosum	Generalised lace-like rash on trunk and 'slapped cheeks' appearance. May be rubella-like. Arthralgia common in adults
Enteric fever	6–10 rose-pink macules over lower chest and abdomen between the 7th and 10th day of fever, fading rapidly whilst new ones appear
Enteroviral rash	Usually a maculo-papular eruption with no other characteristic features
Roseola infantum (exanthem subitum)	Rose-pink rash appearing as the fever of 3–5 days duration rapidly settles
Pityriasis rosea	Oval, reddish-brown papules with fine scales. 'Christmas tree' distribution, preceded by a 'herald patch'
Secondary syphilis	Round, coppery-coloured with superficial scales symmetrically distributed. Palms and soles are involved characteristically
Erythematous rash	
Scarlet fever	Diffuse punctate erythematous rash on 2nd day of fever. Exudative tonsillitis. Circumoral pallor and 'strawberry' tongue. Desquamation of palms and soles during recovery
Kawasaki disease	Persistent fever. Indurated erythema of the extremities. Conjunctivitis, dry red lips, enlarged glands in neck, axillae, groins. Desquamation of palms and soles
Toxic shock syndrome (TSS)	Erythematous patches in an acutely ill, febrile, toxic patient who is hypotensive. Diarrhoea and vomiting common. Desquamation of skin
Erythema marginatum	Fluctuating circinate lesions with clear centres and often a serpiginous outline. Trunk and extremities mostly affected. Associated with rheumatic fever

Condition	*Discriminating features*
Erythema multiforme (Stevens–Johnson syndrome)	Papulo-erythematous lesions of centrifugal distribution. Target lesions and multiple mucosal involvement (eyes, mouth, genitalia) in some. Blisters may develop in severe cases
Toxic epidermal necrolysis (Ritter's disease, scalded-skin syndrome)	Widespread scattered erythematous skin patches which easily separate from underlying tissue on pressure (Nicolsky's sign), leaving a raw area. Flaccid fluid-filled bullae may form
Purpuric or haemorrhagic rash	
Meningococcal disease	Rapidly developing petechial or purpuric lesions with fever. Meningism may be present
Henoch–Schönlein purpura	Usually petechial (may be necrotic and larger) lesion on the distal limbs and buttocks. Arthralgia and abdominal pain. Haematuria may be present
Measles	Severe confluent rash may become haemorrhagic due to leakage of red cells from the capillaries (measles staining)
Viral haemorrhagic fevers (Lassa, Marburg, Ebola, Crimean, Congo, dengue)	Widespread purpuric rash in an ill person with history of recent travel to an endemic country
Typhus fever	In tick typhus, rash may be haemorrhagic; an eschar is often present at the site of tick bite

NB. Thrombocytopenia of any aetiology may cause purpuric rash. Infection-associated causes are: rubella, human immunodeficiency virus (HIV), haemolytic–uraemic syndrome, septicaemia causing disseminated intravascular coagulopathy (DIC).

Papulo-vesicular/pustular rash	
Chickenpox	Papulo-vesicular eruption with an erythematous base. No prodrome.

Condition	*Discriminating features*
	Lesions are of different sizes and stages of evolution because of cropping. Centripetal distribution. Palatal lesions are common
Eczema herpeticum	Vesicular and crusting eruption affecting initially eczematous skin, but later may spread to healthy skin
Hand–foot–mouth disease	Tiny clear vesicles on the dorsum of hands and feet in young children. Discreet, superficial, mouth ulcers
Impetigo	Initially vesicular but rapidly progressing to pustulation and golden-yellow crust formation. The nasolabial area is commonly affected but other areas may become involved through scratching. No palatal lesion
Scabies	Papular, sometimes weepy lesions with greyish burrows. Most common on fingers, wrists, soles and axillae but may be generalised. Very itchy, usually with signs of scratching

Conditions associated with localised involvement of the skin

In these conditions, deeper tissue may or may not be involued.

- Erythematous, tender, indurated lesions: erysipelas, cellulitis, erythema nodosum
- Localised clusters of vesicles: shingles, herpes simplex, impetigo
- Painless nodules: warts, molluscum contagiosum
- Large lesions with pustular or necrotic centres: boils, carbuncles, cutaneous anthrax, orf
- Round scaly patches: ringworm
- Slowly enlarging circular erythema: Lyme disease (erythema chronicum migrans)
- Necrotising/gangrenous lesions: anaerobic cellulitis, gas gangrene, necrotising fasciitis.

SPECIFIC INFECTIONS

Measles

This is caused by the measles virus, a member of the paramyxovirus group. Humans are the only natural hosts.

Epidemiology

Measles occurs throughout the world. It is highly infectious and in unimmunised communities most children will have measles before their 18th birthday, 2-yearly epidemics being common. In countries where childhood vaccination is routine, measles is quite uncommon and older unimmunised persons may remain susceptible through non-exposure. Immunity after natural infection is lifelong. Maternal antibody protects the infant up to the age of 6 months.

Transmission is via respiratory secretion from infected children through inhalation of airborne droplets or by direct contact. Infectivity persists from just before the prodromal period until the 4th day of the rash. The incubation period (IP) is about 10 days (range 7–18 days).

Clinical features

Prodromal illness: abrupt onset of fever, coryza and conjunctivitis, dry cough and miserable appearance, followed on the 2nd or 3rd day by tiny white spots set in red macules on the buccal mucosa, opposite the molar teeth (Koplik's spots). This stage has a duration of 3–4 days

Rash: on about the 4th day, dusky red maculo-papules appear behind the ears and on the face, and spread downwards slowly to reach the lower limbs 3 days later. Fever subsides soon afterwards. Rash on the face and trunk often becomes confluent before fading over the next 2–3 days.

Complications

- Secondary bacterial otitis media and bronchopneumonia, febrile convulsion
- Post-infective encephalitis (1/1000 cases). Subacute sclerosing panencephalitis — a late sequela (1/1000,000 cases). Progressive encephalitis in immunocompromised
- Severe measles in malnourished children in tropics: prolonged rash, diarrhoea and secondary bacterial complications.

Diagnosis

Confirmation is by demonstration of immunoglobulin M (IgM) antibody in blood or saliva or four-fold rise of immunoglobulin antibody during convalescence. Demonstration of the virus in nasopharyngeal secretion is the best means of diagnosis in immunocompromised children who may not produce a typical antibody response.

Treatment

In hospitals, children should be isolated in single rooms; in communities, kept away from school until after 4 days of rash. Most cases require symptomatic care only. Co-amoxiclav is appropriate for secondary bacterial infections.

Prevention

Measles vaccine: in Britain and many other countries, live attenuated measles vaccine is given routinely (combined with mumps and rubella vaccines (MMR)) during the 2nd year of life. The vaccine gives $>$95 per cent protection, possibly for life. About 5 per cent of recipients develop a short-lasting febrile illness, sometimes with a transient rash after vaccination, but other side effects are very rare. It is contraindicated in the immunocompromised. In institutional outbreaks, the vaccine may provide protection to contacts if given within 72 hours of exposure. Encephalitis has been reported very rarely

Normal immunoglobulin: this will often prevent or attenuate illness if given within 6 days of exposure and is recommended for exposed susceptible children who are immunocompromised or for other household and institutional contacts.

Prognosis

Measles is rarely fatal in previously healthy children but it is an important cause of morbidity and mortality in malnourished children in developing countries.

Rubella (German measles)

The aetiological agent is rubella virus, which is a togavirus. Humans are the only hosts.

Epidemiology

Rubella occurs worldwide – both sporadically and in outbreaks – and many infections are subclinical. It is highly infectious and in unvaccinated communities around 80 per cent of adults would have had prior infections. Due to routine childhood immunisation, rubella incidence has declined dramatically in many Western countries. Immunity is lifelong.

Transmission is via airborne droplets of nasopharyngeal secretion of infected people 5 days before to 5 days after their rash develops. Transplacentally infected babies are often chronically infected, excreting virus in throat and urine (for up to 1 year) and are potential sources of infection to others. The IP is 16–18 days.

Clinical features

Prodromal illness: usually absent in young children but adults often have a short period (up to 2 days) of mild fever and malaise. Enlarged, tender glands appear behind the neck

Rash: discrete, rose-pink macules appear first on the face and neck, then spread to trunk and limbs within 24 hours. Mild conjunctivitis may develop. The rash then rapidly fades in order of appearance

Polyarthralgia: may affect finger joints and other large joints bilaterally in adults, lasting over several weeks.

Complications and prognosis

Rubella is usually very benign in the acquired form, with two rare complications: (i) post-infectious encephalitis (1 in 5000 cases); and (ii) immune thrombocytopenia.

Congenital rubella syndrome

When infection occurs in early pregnancy, fetal death may occur or the baby may be born with a number of defects: cataract, microcephaly, perceptive deafness, patent ductus arteriosus or septal defect, hepatosplenomegaly, thrombocytopenia, mental retardation. Insulin-dependent diabetes may develop. The risks and severity of fetal damage vary with the time of infection in pregnancy: highest incidence (up to 80 per cent) and major defects during the first 4 weeks, falling to 25 per cent during 8–12 weeks and to around 10 per cent during 12–16 weeks (mostly deafness, detected only later due to child's learning difficulty). Fetal damage is rare thereafter.

Diagnosis

Serological confirmation (presence of IgM antibody or four-fold rise of IgG antibody during convalescence) is mandatory during early pregnancy. Parvovirus in adults often produces a similar illness. Congenital rubella syndrome is diagnosed by demonstration of IgM antibody and/or virus isolation from throat or urine.

Treatment

There is no specific treatment. Termination of pregnancy is a reasonable option in maternal rubella, depending on the stage of pregnancy and maternal wish.

Prevention

• Normal immunoglobulin following exposure does not prevent infection

• In Britain and many other countries, children are now routinely given highly effective (>95 per cent seroconversion) live rubella vaccine RA 27/3 (as MMR) during their 2nd year of life. Immunity is long lasting and, although subclinical natural infection can occur in vaccinated individuals, fetal damage occurs very rarely. The vaccine is contraindicated in the immunocompromised and during pregnancy (for fear of fetal damage – not reported although fetal infection may occur).

Erythema infectiosum

The cause of this illness, parvovirus B19, belongs to the family of Parvoviridae. Humans are the only hosts.

Epidemiology

• Occurrence is worldwide – both sporadically and in epidemics. Infection is common in children and is often asymptomatic. Immunity is lifelong

• Transmission is through inhaled droplets of or contact with infected respiratory secretions, or transplacental in the case of fetal infection. IP is variable (4–20 days).

Clinical features

• A short prodrome of malaise may occur in older children but usually the first sign of illness is a striking erythema of both cheeks (slapped cheek appearance) followed a day or two later by a maculo-papular rash on the trunk and limbs which often assumes a lace-like appearance. This may reappear after fading, particularly on exposure to sunlight. Fever is unusual. In adults, the rash is often atypical and may be rubella-like

• Sore throat and polyarthralgia/polyarthritis and mild generalised lymphadenopathy are not uncommon in adults.

Complications

• Transient aplastic crisis (in patients with haemolytic anaemia) and severe chronic anaemia (in the immunocompromised)

- Infection in pregnancy can rarely cause hydrops foetalis and fetal death, but congenital anomalies do not occur.

Diagnosis

Confirmation is by demonstration of IgM antibodies or four-fold rise of IgG antibodies during convalescence. Demonstration of viral antigen in the blood of patients with severe anaemia or in aborted fetal tissue by DNA examination is helpful.

Treatment

Treatment is entirely symptomatic. Efficacy of normal immunoglobulin in preventing infection in contacts is unknown.

Roseola infantum (exanthem subitum)

Epidemiology

This acute febrile illness of worldwide distribution is caused by human herpesvirus 6 (HHV-6). Almost all cases occur in pre-school children – mostly in <2 year olds. Transmission is through contact with the respiratory secretions of an infectious person.

Clinical features and diagnosis

After an IP of approximately 10 days, high fever develops abruptly and usually lasts for 3–5 days. As fever subsides, a maculo-papular rash appears over the trunk, and moves to the rest of the body before fading rapidly. Febrile convulsion at the start of the illness is the only important problem.

Several other viruses produce a roseola-like rash, principally rubella and enteroviruses.

Pityriasis rosea

Epidemiology

This is a common dermatosis of possible infective aetiology and affects mainly young adults.

Clinical features

The lesions are oval, reddish brown papules with fine scales, most prominent on the trunk and particularly the back, and often have a 'Christmas tree' shaped distribution with lesions following the sloping ribs. Careful search will usually reveal a single large lesion somewhere on the body which has appeared before the generalised rash (herald patch). Constitutional symptoms are rare.

Differential diagnosis

The differentiation is from secondary syphilis, in which the early lesions are often round, generalised and symmetrical with superficial scales, but the palms and soles are also affected.

Treatment and prognosis

There is no treatment. The lesions usually disappear after 2–3 weeks.

Scarlet fever, erysipelas, cellulitis, streptococcal pyoderma and impetigo

These conditions are all skin manifestations of *Streptococcus pyogenes* infections.

SCARLET FEVER

Epidemiology and pathogenesis

Scarlet fever is a complication of streptococcal infection, usually of the throat but less commonly of other sites (e.g. burns, chickenpox lesions, surgical wounds). The characteristic rash is caused by streptococcal erythrogenic toxin which stimulates immunity, so second attacks of scarlet fever at rare, even if recurrent attacks of streptococcal infections are common. The IP is 1–3 days.

Clinical features

- After a prodromal period (1–2 days) of fever and sore throat, rash appears: fine red spots on a background of diffuse pinkness, initially on the neck and chest, spreading downwards to affect the rest of the body. The cheeks are flushed with pallor around the mouth (circumoral pallor)
- Strawberry tongue: the papillae of the tongue become inflamed and red, protruding through a white coating on the tongue
- Desquamation: towards the end of the week, the skin peels, with a fine desquamation on the trunk and large flakes on the hands and feet.

Diagnosis

The presence of the typical rash in a patient with streptococcal pharyngitis is diagnostic. Desquamation is also seen in Kawasaki disease and toxic shock syndrome. Antistreptolysin O (ASO) titre is raised during convalescence.

Complications

These are as in streptococcal throat infections (see p. 52). Cardiotoxicity used to be a complication of scarlet fever but is nowadays very rare.

ERYSIPELAS

Epidemiology and pathogenesis

In contrast to staphylococcal infection of the skin, which often generates localised pus-producing lesions, *S. pyogenes* tends to cause spreading lesions due to the production of a variety of extracellular toxins which destroy fibrin, cellular proteins and hyalnuronic acid. This facilitates the spread of infection through the tissues – resulting in conditions like erysipelas, cellulitis and, rarely, streptococcal fasciitis. Erysipelas may occur at any age, although it is more common in the over-40s age group. It is caused by streptococcal toxin infiltrating the skin from a small primary focus, which may be invisible.

Clinical features

There is a florid erythema which is tender, slightly raised and with an advancing, clear-cut edge. On the face, a common site, the rash is often on both sides of the nose, which itself is red and swollen. In late cases, blisters may appear. The condition can recur repeatedly, particularly in the legs, around chronic ulcers and may cause lymphatic damage leading to lymphoedema.

Diagnosis

This is usually clinical unless blisters are present, facilitating the isolation of streptococci. There is neutrophilic leucocytosis and a rise in ASO titre during convalescence.

OTHER STREPTOCOCCAL SKIN INFECTIONS

Cellulitis

Cellulitis is a subcutaneous infection due to *S. pyogenes* (less commonly *Staphylococcus aureus* and *Clostridium welchii*) with lymphatic spread. As in erysipelas, the portal of entry may be inapparent. There is a sheet of raised, tender erythema without a clear-cut margin but with projections along the lines of lymphatics. This may be complicated by necrotising fasciitis.

Streptococcal pyoderma

Streptococcus pyogenes commonly infects wounds, cuts, burns and abrasions. In mild form, infection delays healing with excessive redness around the wound and excessive serous discharge. In more obvious infection, the discharge forms a crust under which there are pockets of pus. Pyoderma is endemic in children in the tropics. In temperate climates, it can occur in outbreaks in burns units and penal institutions, in meat workers and military recruits.

Impetigo

This is a stage of infection beyond pyoderma. There is a spreading vesicular rash, as streptococci invade apparently healthy skin. The rash is localised, often around the nose or mouth. The vesicles break easily, exuding seropurulent fluid and forming thick, golden-yellow crusts. The lesions may also be infected by *S. aureus* and sometimes this may cause tense, pus-filled bullae (impetigo bullosa). Impetigo is predominantly a disease of children.

Necrotising fasiitis

See p. 171.

Treatment

Penicillin is the drug of choice, the route of administration depending on the severity of infection. With erysipelas, cellulitis and severe scarlet fever, it is worth starting with intravenous (IV) therapy. Erythromycin is a suitable alternative for penicillin-allergic patients. In pyoderma and impetigo, the penicillin chosen should also cover penicillinase-producing staphylococci. Topical antibiotics are ineffective. Local disinfectants and cleanliness will help to control or prevent outbreaks of streptococcal skin infections.

Boils, carbuncles, toxic shock syndrome and toxic epidermal necrolysis

These conditions are all skin manifestations of *S. aureus* infections.

BOILS AND CARBUNCLES

These are among the most common soft tissue infections and are found in any part of the body. Boils (furuncles) result from infection of hair follicles or sebaceous glands, presenting as a firm, tender nodule which later develops a central pustule. Carbuncles are large cutaneous abscesses with deeper subcutaneous involvement and multiple pustular heads. These are particularly common in diabetic, immunocompromised and malnourished individuals. Local and distant spread and septicaemia may complicate even a simple boil.

Spontaneous or surgical drainage of pus usually cures a boil. Antibiotics (flucloxacillin) are indicated for boils on vulnerable sites (e.g. the face) or when signs of spread are present, and for carbuncles.

TOXIC SHOCK SYNDROME

Epidemiology

This is a severe illness caused by toxin-producing staphylococci. It is particularly associated with vaginal infection during tampon use, but may follow infection at any other site.

Clinical features

- Presentation is acute, with high fever, watery diarrhoea, vomiting and muscle pains, followed by hypotension and possibly shock
- Sunburn-like erythematous rash and hyperaemic mucosa (vaginal, buccal, conjunctival). Skin desquamates after 1–2 weeks
- Evidence of damage to two or more organ systems, e.g. muscles (raised creatine kinase (CK)), liver (raised serum glutamic pyruvic transaminase (SGPT)), renal (pyuria, raised urea/creatinine), central nervous system (CNS) (disorientation, drowsiness, focal signs).

Diagnosis

Other infections which may cause a similar clinical picture need exclusion (see p. 142). Staphylococci should be looked for in different body sites, and isolates should be tested for phage group and toxigenesis.

Treatment and prevention

Prompt therapy with anti-staphylococcal antibiotics, eradications of focus (removal of tampons, drainage of pus) and general support are essential. The prognosis is good. The risk of TSS can be reduced by frequent changing of tampons and use of less absorbent tampons.

SCALDED-SKIN SYNDROME (RITTER'S DISEASE, TOXIC EPIDERMAL NECROLYSIS, LYELL'S DISEASE)

This is caused by exotoxin-producing *S. aureus* of phage group II. There is spreading damage to the superficial layers of the epidermis. The disease presents with high fever and widespread erythema over the body. Large superficial blisters appear on the erythematous areas, which rupture and strip away like wet tissue paper leaving a raw, red surface. Drying and healing occur within a few days. Staphylococci are isolated from the skin lesions.

Oral or IV flucloxacillin, depending on the severity, is the drug of choice.

Kawasaki disease (mucocutaneous lymph-node syndrome)

Epidemiology

The aetiology of this condition is unknown. Search for an infective agent has so far proved unsuccessful. An abnormal immunological response to a variety of microbial antigens including streptococci, rickettsiae, viruses and house dust mites has been postulated. The greatest prevalence is in Japan but cases have been reported worldwide, including Britain. It occurs predominantly in children below 5 years of age.

Clinical features and diagnosis

- Five main features are recognised: fever for more than 5 days, non-purulent bilateral conjunctivitis, reddened lips and oral mucosa, raw red tongue, erythema of limbs including palms and soles, and oedema may be present (later periungual or more generalised desquamation occurs), cervical adenopathy
- Persistent leucocytosis, high erythrocyte sedimentation rate (ESR) or plasma viscosity and thrombocytosis (2nd week onwards) are characteristic laboratory features
- Up to 20 per cent of cases develop coronary aneurysms due to coronary arteritis during the later phase of illness, which may last several weeks. Long-term prognosis of recovered cases is not known. Other organs of the body, e.g. liver, lung or joints, may be affected during the acute stage.

Diagnosis and treatment

Diagnosis rests on suggestive clinical and laboratory features. Differential diagnoses includes scarlet fever, TSS, erythema multiforme, juvenile Still's disease or infantile polyarteritis.

High-dose IV gammaglobulin and aspirin during the early stage reduce the risk of coronary aneurysm formation. Follow-up should be prolonged.

Prognosis

Although generally benign, about 1 per cent of cases end fatally, mostly from cardiac complications.

Erythema multiforme

Epidemiology and pathogenesis

An abnormal immunological reaction of the skin to exogenous stimulants is probably the underlying pathogenesis of this condition, but the exact mechanism of disease production is not clear. There is subepidermal damage to the skin. It may follow *Mycoplasma* or herpes simplex virus (HSV) infection or drug administration (particularly sulphonamides and penicillin), but often no precipitating cause can be identified. Children and young adults are mainly affected.

Clinical features

- Patients often have a short prodromal period of malaise and fever. A bright papulo-erythematous rash develops with lesions of varying sizes and shapes, affecting characteristically the extremities including palms and soles, often symmetrically. Other areas of the body may also be affected. The lesions may have a raised peripheral ring with paler centre (target lesion) (see Plate 3, facing p. 180). Bulla formation may occur
- In some cases, there is inflammation of mucosal surfaces (eyes, mouth and genitalia), when Stevens–Johnson syndrome is diagnosed. High fever and pneumonitis may complicate the picture.

Diagnosis and treatment

Diagnosis is clinical and requires exclusion of other erythematous and blistering diseases.

Attention to hydration, nutrition and treatment of secondary bacterial infection is important. Corticosteroids may help in severe cases. Recurrences may continue and if there is a clear relationship with recurrent HSV infections, acyclovir suppression therapy is often helpful.

Henoch–Schönlein purpura

This is an immunologically mediated vasculitic condition, often triggered by a preceding *S. pyogenes* infection but in other cases the aetiology remains unknown.

Children and young adults are usually affected.

Clinical features

The illness is characterised by the triad of purpuric rash (around the elbows, ankles and over the buttocks), abdominal pain, and arthralgia or arthritis. Fever and malaise are common but the patient is not acutely ill.

Associated glomerulonephritis (in 30 per cent of cases) and, rarely, gut bleeding or intussusception may complicate the clinical picture.

Prognosis

Prognosis is usually good with natural remission within a few weeks but adults may develop chronic glomerulonephritis and hypertension.

Treatment

Treatment is symptomatic, e.g. analgesia for joint pain. Corticosteroids may help abdominal symptoms but do not prevent progression of renal disease. Immunosuppression therapy may be needed for persistent glomerulonephritis.

Herpes simplex virus infection

HSV belongs to the large group of Herpesviridae family which are DNA-containing enveloped viruses, distributed widely throughout the animal kingdom. Eight human herpesviruses are recognised – HSV-1, HSV-2, varicella zoster virus, cytomegalovirus (CMV), Ebstein–Barr virus (EBV) and human herpesvirus 6, 7 and 8. Monkey herpes virus, herpes B virus (herpesvirus simiae), is a rare cause of encephalitis in humans. HSV has two types: HSV-1 is the cause of most oral, conjunctival and cutaneous infections and HSV-2 is the cause of most genital infections.

Epidemiology

- HSV infections are widespread worldwide. Transmission is through direct contact with infected secretions, either oral or genital. Babies in the first few months of life are usually immune because of transferred maternal antibodies; thereafter the rate of infection (HSV-1) rises rapidly during early childhood and by adulthood most have been infected
- HSV-2 is rare before adolescence and usually begins with sexual activity, and prevalence is high in sexually promiscuous groups
- The IP is around 5 days (range 2–12 days).

Pathogenesis and clinical features

In susceptibles, skin and mucous membranes are the primary sites of multiplication, with further multiplication in regional lymph nodes, and viraemia in some cases. Virus also travels to local sensory ganglia via sensory nerves and may lie dormant for life after subsidence of primary infection, with reactivation to recurrent forms of HSV disease.

Primary infections

These are often asymptomatic (about 90 per cent) but, when clinically apparent, may manifest in several forms.

• *Herpetic gingivo-stomatitis:* most common; pre-school children mostly affected. Febrile illness with multiple shallow, slough-covered ulcers in the mouth and vesicles on and around the lips. Cervical glands are enlarged. Recovery in a week or so

• *Keratoconjunctivitis:* a unilateral follicular conjunctivitis with characteristic herpes vesicles on the eyelids, with or without keratitis

• *Primary ano-genital herpes:* mainly in adults; sexually transmitted. There are painful vesicles and ulcers on the vulva, vaginal mucosa and cervix, or on the glans, prepuce and shaft of the penis. Lesions may spread to the adjacent areas of the perineum, buttocks and thighs. Healing usually occurs within 2–4 weeks. Sacral radiculitis may complicate the clinical picture. Fever is common

• *Skin herpes:* in susceptible individuals, a breach in the skin may become infected and result in a cluster of vesicles. When a finger is the site involved, a painful paronchia may result (herpetic whitlow). In patients with eczema, primary herpes infection of the skin may cause a more serious illness. Extensive areas may be affected with constitutional disturbance, and may end fatally (eczema herpeticum)

• *Neonatal herpes:* infection can arise either *in utero* from maternal viraemia from a primary infection during pregnancy, or during passage through the mother's infected birth canal (more common). Most infections are due to HSV-2. The risk of fetal infection is much higher in mothers with primary genital disease during delivery. Clinical presentation depends on the timing of the infection: early *in utero* infection may lead to abortion, premature labour, retarded growth and nervous system abnormalities with microcephaly and choroidoretinitis. In infections nearer full term, the baby at birth may have signs of generalised infection with features of hepatitis, pneumonitis, CNS involvement and a vesicular rash. Severe disease is rare in babies born of mothers with recurrent genital herpes (due to the protective effect of maternal antibodies)

• *Herpes encephalitis:* herpes encephalitis is the most common form of severe sporadic primary encephalitis in the UK (annual incidence: 50–100 cases) and may occur during primary infection (rare) or due to reactivation of latent HSV-1 infections (more common). The rare HSV-2 CNS infection in adults usually causes aseptic meningitis.

Recurrent infections

The virus lying dormant in the regional sensory nerve ganglia may reactivate from time to time, causing lesions on skin or mucous membranes (cold sore, whitlow, genital herpes, dentritic keratitis, skin herpes). A prodrome of burning, tingling and itching of several hours duration is common in cold sores. Intercurrent illnesses, sunlight and menstruation are common triggering factors. Individuals with depressed cell-mediated immunity (CMI) are prone to frequent or persistent severe recurrences, usually of oro-facial and ano-genital types. Disseminated infection with deeper organ involvement may occur (liver, oesophagus, brain).

Diagnosis

Confirmation is by the demonstration of virus in vesicle fluid or ulcer smear by electron microscopy (but this will not distinguish it from other herpesviruses), or by culture. Demonstration of a four-fold rise of antibodies is useful in documenting a primary infection, and measurement of IgM HSV antibodies in infants may be helpful in the diagnosis of neonatal herpes.

Treatment

- Acyclovir is effective against HSV, and definite indications are encephalitis, eczema herpeticum, primary skin and genital infections, severe primary stomatitis, infections in neonates, the immunocompromised, and of the eye. Ocular infections are treated with a 3 per cent ophthalmic formulation (with systemic administration if primary)
- Frequent labial or genital recurrences are treated with continued acyclovir suppressant therapy or intermittent self-treatment by patient at the earliest warning sign. A 5 per cent topical formulation is less effective for this purpose.

Prevention

Patients and carers with active lesions should cover these or avoid predisposed individuals. Prevention of HSV transmission to the newborn is difficult to achieve. If there is florid maternal genital infection (suggestive of primary infection) at the time of labour, elective caesarean section is advisable and this should be carried out within 6 hours of membrane rupture. Vaginal delivery otherwise does not carry any additional risk, even if the mother is known to suffer from recurrent genital herpes.

Chickenpox and herpes zoster

These common but clinically dissimilar diseases are caused by a single serotype herpesvirus – varicella zoster virus (VZV). Chickenpox is the manifestation of primary infection with VZV, whereas herpes zoster is caused by reactivation of latent VZV. Humans are the only hosts.

CHICKENPOX

Epidemiology

Chickenpox is primarily a childhood disease of worldwide prevalence. In urban communities most adults would have had chickenpox, but in isolated communities many adults may remain susceptible. Transmission is via airborne inhalation of respiratory secretion from chickenpox patients, or by direct contact with skin lesions (chickenpox and zoster). Immunity is lifelong

The IP is usually 13–17 days (range 10–21 days).

Clinical features

Prodromal period: this is usually absent in children, but adults may have fever and malaise for 1–2 days

Rash: this is usually maculo-papular, and rapidly progresses through vesicular, pustular and crusting stages over a period of 3–4 days. New lesions continue to appear. Crusts separate after about 1 week, leaving a shallow pit which soon disappears. Lesions are more profuse on the head and trunk and on areas of irritation (sunburn, napkin dermatitis). Mucosal lesions are not uncommon. There is a great variation in the profuseness of rash, it being more common in adults and the immunocompromised

Infectious period: the patient is infectious from 1–2 days before the appearance of rash to 5 days after. In immunocompromised patients, the skin lesions may remain infectious for a longer period.

Complications

- Secondary skin sepsis
- Pneumonitis: in immunocompromised patients and previously healthy adults; smokers and pregnant women are more susceptible. In severe cases, respiratory failure develops due to extensive bilateral alveolitis. Radiologically there are discreet opacities scattered throughout both lungs, some of which may calcify after recovery (Fig. 10.1)
- Post-infectious cerebellar encephalitis (1/6000 cases): usually benign, presenting only as ataxia
- Chickenpox in the immunocompromised: individuals with depressed CMI often have a severe, more prolonged illness, with complicating pneumonia or encephalitis and cropping of new lesions even after recovery

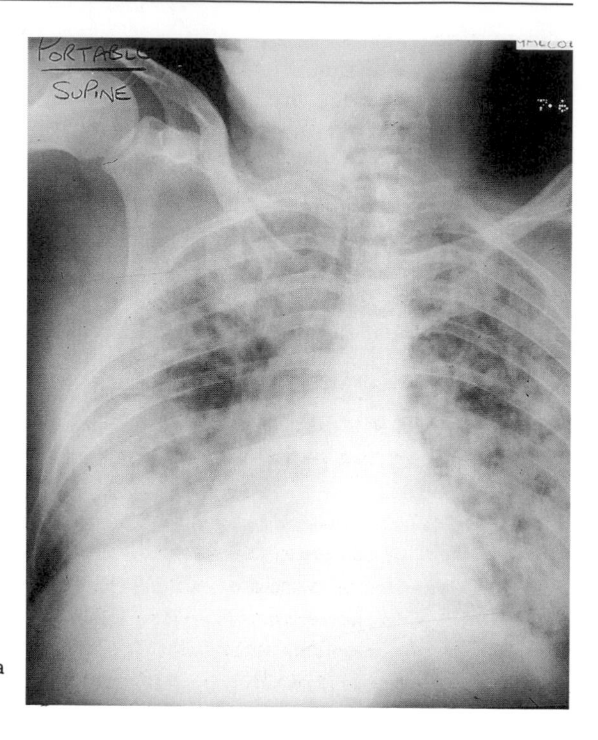

Fig. 10.1. Primary varicella pneumonitis.

- Chickenpox in pregnancy and the newborn: fetal malformations may occur in up to 3 per cent of first-trimester infections, e.g. low birth weight, short limbs, eye/brain damage, zoster-like rash (congenital varicella syndrome). Newborns up to 4 weeks after birth may develop severe disease if the mother has rash within 7 days of birth, or when contact is postnatal and mother is non-immune.

Diagnosis

In doubtful cases, either serology (four-fold rise of antibody) or demonstration of virus in vesicle fluid (electron microscopy (EM), culture) are diagnostic. Multinucleated giant cells are usually seen in Giemsa-stained scrapings from the base of a lesion (also seen in HSV lesions).

Treatment

Patients with respiratory symptoms should be given IV acyclovir. All adults, including pregnant women, and all immunocompromised patients with signs of continuing active infection (fever, continuing cropping of lesion) should receive acyclovir (IV or oral). Acyclovir has no teratogenic effect.

Prevention

- It is customary to exclude children from school until the rash has resolved, but the usefulness of this practice is questionable as most children will develop the illness eventually. Patients in hospital should be isolated to protect other patients vulnerable to severe chickenpox
- Varicella zoster immunoglobulin often modifies the illness if given within 10 days after exposure to chickenpox/zoster, and should be given to the following groups: susceptible immunosuppressed patients and pregnant women, newborns prone to develop severe disease (see Complications above).

HERPES ZOSTER

Epidemiology and pathogenesis

Zoster results from reactivation of latent VZV in a dorsal root or extramedullary cranial nerve ganglion. The virus travels down the sensory nerve to the area of skin supplied by the nerve and produces the typical skin lesions. Depressed CMI, due to advancing age, disease or drugs, promotes reactivation. Zoster does not result from contact with another case of zoster or chickenpox.

The condition is relatively uncommon in persons below 40 years old but the rate rises progressively after 50 years of age and persons over 80 years old have a 10 per cent chance of developing zoster each year. Zoster in very young children is probably related to less efficient development of CMI at the time of the primary VZV infection, either *in utero* or postnatal.

Clinical features

The following stages are recognised.

Prodromal pain: 2–3 days duration on average, but may be longer

Rash: red patches with vesicles developing on them, progressing to pustules which scab and then separate. New lesions occur for 3–5 days. Scabbing and separation are completed within 2–4 weeks. A single dermatome is mainly involved, with lesions commonly spreading on to adjoining dermatomes on the same side because of overlapping innervation. Dorsolumbar dermatomes are most frequently involved ($>$ 50 per cent), followed by trigeminal (ophthalmic is most common), cervical and sacral. Extremities are least commonly affected

Acute phase pain: common during rash evolution.

Complications

- Secondary skin sepsis
- Ocular: conjunctivitis, keratitis, uveitis, retinal necrosis, scarring of lids
- Cutaneous or visceral dissemination in the immunocompromised
- Paralytic zoster, due to involvement of motor nerves. Examples are Ramsay Hunt syndrome (painful eruption in and around the ear, ipsilateral lower motor neuron (LMN) VII nerve palsy with or without vestibular disturbances) (see Plate 4, facing p. 180), external ophthalmoplegia, bladder disturbances and weakness of limb muscles
- Meningo-encephalitis and myelitis (rare)
- Post-herpetic neuralgia (PHN): this is best defined as persisting dermatomal pain after healing. Overall its incidence is 9–15 per cent but reaches 50 per cent in persons over 60 years old. A small but significant number will continue to have distressing pain beyond 6 months. PHN is variable in severity, type and quality.

Treatment

- Adequate analgesia is important in all stages
- Antiviral drugs (acyclovir, famiclovir, valaciclovir) shorten the duration of acute illness, prevent dissemination and probably shorten the duration of pain. The patients most likely to benefit are:
  - immunocompetent patients >60 years old (more likely to develop severe acute disease and PHN); must be given within 72 hours to be of benefit
  - patients with ophthalmic zoster, within 72 hours of onset
  - all immunocompromised patients with active disease at any stage
- In established PHN, amitriptylene, topical cool spray and trans-cutaneous nerve stimulation are often helpful.

Cutaneous infestations and insect bites

SCABIES

This is a skin infection caused by the scabies mite, *Sarcoptes scabiei*. The disease varies in incidence over the years for uncertain reasons.

Epidemiology and pathogenesis

Scabies is spread by close skin contact, typically by hand holding or sharing beds. Contrary to popular belief, scabies is not spread by clothes or bedding. The mite burrows into the skin to the boundary between the strata corneum and the granulosum. The mite stays in the burrow, laying eggs which hatch into larvae to become adult mites in 10–

14 days. The females are fertilised by male mites when they have made new burrows.

Clinical features

- The rash and itching of scabies are caused by sensitisation to the mite proteins, so symptoms may not start until infection has been present for 2 or 3 weeks, and may remain for the same time after the mites have been killed by insecticides. The itching is worse at night. Enquiry may find that another person in the family, or a close friend, also has itching
- The rash is usually symmetrical, on the fingers (especially on the webs), wrists, penis, around the waist and on the buttocks. Normally the head and back are spared. The most common lesions are papular, with scratch marks, but the diagnostic lesion is the burrow, a pin-head-sized blister with a short line.

Diagnosis

The diagnosis of scabies can be confirmed by lifting the mite out of a burrow with a needle and examining it under a low-power microscope. Skin scrapings from a burrow or an unexcoriated papule will gather eggs and a mite to be examined in the same way.

Complications

Scabies in children can be complicated by secondary infection, particularly by group A streptococci. In tropical countries, there have been outbreaks of glomerulonephritis caused by this complication. In contrast to the relatively few lesions on a person who washes frequently, people who are unable to care for themselves may have extensive lesions with crusting – this form is sometimes referred to as Norwegian scabies and is more contagious than the usual case (it is seen particularly in immunodeficient and senile patients).

Treatment

- Scabies is treated with topical insecticides. Lindane or malathion lotion should be applied to the whole body, excepting the head, paying attention to the hands and feet. Lindane is not advised for pregnant and breast-feeding women
- The whole family and sexual partners should be treated at the same time. A single application should be sufficient. There is no need to wash clothes or bedding in any special way, because these articles do not harbour mites. Patients should be told that the itching may persist for some time, but can be relieved with calamine lotion.

HEAD LICE

Head lice are fairly common, being adapted to modern human society. The lice, *Pediculus humanus capitis*, live close to the scalp and lay their eggs at the roots of hairs. The egg capsules, nits, are seen attached to hairs. Head lice cause itching, but no serious disease. They spread by close contact within families. Since schoolchildren are the people in whom the diagnosis is most commonly made, there is a mistaken belief that transmission and its prevention depends on schools. However, the control is by treating the families of cases, with malathion, carbaryl or permethrin lotions and shampoos. As insecticide resistance can occur, it is a common policy to rotate through the various insecticides, using only one type in a district at a time.

BODY LICE

While head lice can infect people regardless of their hygiene, body lice, *Pediculus humanus corporis*, are associated with a lack of cleanliness. This infestation is more common in people without homes, and those in refugee camps. The bites cause itching, the main symptom. The important health factor is that, in refugee and war conditions, lice can be vectors of typhus and relapsing fever. In addition to treating patients with insecticide lotions, clothing and bedding should be washed thoroughly and hot ironed, and washing facilities should be improved.

Crab lice, *Pthirus pubis*, are usually transmitted during sexual intercourse. Although itching may be present, doctors are more likely to find crab lice either as an incidental observation or in association with other sexually transmitted infections.

INSECT BITES

In addition to transmitting a variety of diseases, insect bites can cause appreciable morbidity. Humans build up tolerance to bites that they encounter frequently, but before this happy state is reached, bites can cause discomfort, itching, urticaria and papular or vesicular skin lesions. The vesicular lesions are prone to secondary infection in unhygienic conditions. The main medical task with insect bites is to distinguish them from other skin disorders and to advise on how to avoid further bites. When lesions are confined to areas of skin that are not covered by clothing, insect bites should be considered.

In Britain, flea bites on the arms and legs result in a condition called papular urticaria. The fleas have usually come from pet dogs or cats and have hatched from eggs lying in carpets or grass (in summertime). The patient may not suspect flea bites if the fleas have come from another family's pet. The condition mostly affects children between the ages of 2 and 7 years, and it is often seasonal. It is controlled by disinfesting pet animals and carpets.

CUTANEOUS MYIASIS

This is caused by the maggots (larvae) of different types of flies. The larvae require a period of maturation underneath the skin of warm-blooded animals – usually involving wild rodents, but humans can become the host in appropriate circumstances. Tumbu flies (*Cordylobia anthropophaga*) and human botflies are commonly involved. (In Britain, the occasional cases are seen in travellers.) In west and central Africa tumbu flies are widely prevalent, depositing eggs on sandy soils or clothing left to dry. On contact with warm skin, the eggs rapidly mature into invasive larvae, which penetrate the skin. After maturing for about 2 weeks they fall out to the ground and develop into mature flies. Human botflies are seen in the forested areas of tropical America and produce a similar illness although the larvae take 2–3 months to mature. Botfly eggs are transported onto animal skin by blood-sucking insects.

Presentation is usually as 'boils', which are often multiple. The centre of the 'boil' has a black mark which is the opening of the larva's spiracle through which it breathes. Extraction of the larva is best achieved by applying an oily substance to the central opening. This suffocates the larva and it begins to wriggle out, when it can be removed easily with forceps.

Erythema nodosum

In this condition, there are raised, shiny, tender red areas on the shins, and less commonly on the forearms or elsewhere. As the lesions fade, they may resemble bruises. The patient may be febrile, have a raised ESR and have polyarthralgia. The condition may last for several weeks. Children and young adults are most commonly affected.

Causes

Infections: *S. pyogenes*, tuberculosis, leprosy, *Yersinia*, *Chlamydia*, *Salmonella*, *Campylobacter*, histoplasmosis, toxoplasmosis, glandular fever
Drugs: sulphonamides, oral contraceptives, dapsone
Other: sarcoidosis.

Diagnosis and treatment

Diagnosis and treatment consist of identification of the precipitating factor (often not found) and its treatment or removal. Otherwise, treatment is symptomatic. Corticosteroids may alleviate symptoms rapidly but are rarely necessary and should be used with caution where the aetiology is uncertain.

Warts, molluscum contagiosum, orf and paravaccinia

WARTS

These are caused by human papilloma viruses (HPV) through inoculation of the skin or mucous membrane (squamous).

- Warts vary from painless nodules or papular warts to flat plantar warts with inward growth; both are extremely common in children. In moist genital areas, warts are often larger, fleshy and cauliflower-like (particularly in the immunocompromised), and may involve the vaginal and anal mucosa. Laryngeal warts may occur in children
- Transmission is through direct contact, or via contaminated floor surfaces. Genital warts are sexually transmitted
- In children, virtually all warts disappear spontaneously within 3 years, and many within a few months. Cryotherapy or local application of podophyllin expediates resolution. Surgery may be necessary in large, genital growths.

MOLLUSCUM CONTAGIOSUM

This is a common infection of children worldwide, and is caused by a pox virus. Transmission is through direct contact or via fomite.

- Lesions are usually multiple (may be numerous in immunocompromised). The ano-rectal region, trunk and face are common sites. The individual lesions are discreet, pearly papules with central umbilication
- Spontaneous resolution occurs normally within 1 year in the immunocompetent, but lesions often persist and become larger in the immunocompromised
- The clinical appearance is usually diagnostic. Light microscopic examination of the cheesy material expressed from a lesion will reveal characteristic molluscum bodies (intracytoplasmic inclusions). EM will reveal the pox virus, but is rarely needed. Cryotherapy is helpful in persistent cases.

ORF (CONTAGIOUS PUSTULAR DERMATITIS) AND PARAVACCINIA (MILKER'S NODULE)

These are occupational diseases of humans caused by viruses belonging to the family Poxviraediae which, in nature, infects sheep or goats (causing vesicular stomatitis) and cattle (causing lesions on udders: paravaccinia). Transmission is through contact with mucosal secretion.

- In orf there is usually a solitary papule on a finger or other exposed areas of skin. This slowly enlarges, forming a multiloculated blister with little pain and slight surrounding erythema
- Paravaccinia lesions are usually multiple, bluish papules without pustulation
- The resolution of both is spontaneous within a few weeks. The diagnosis can be confirmed by EM demonstration of the virus in material from lesions (see Plate 5, between pp. 180 and 181).

Ringworm (dermatophytosis) and pityriasis versicolor

RINGWORM

The skin infections caused by the dermatophyte fungi, the species *Microsporum*, *Trichophyton* and *Epidermophyton*, are clinically termed tinea capitis, tinea corporis, tinea pedis, etc., according to the part of the body affected.

Epidemiology

Infection is acquired by contact with the skin scales of other affected individuals or occasionally those of animals. Communal bathing areas or shower rooms facilitate transmission.

Clinical features

- Tinea corporis starts as a red, scaling area which usually enlarges with central healing. More acute lesions are itchy and may be vesicular. Groin and feet infections are usually more chronic and the skin becomes macerated and uncomfortable
- Infected nails are dull, brittle and discoloured, thickened and distorted
- Scalp infections are most common in children. They start as scaling papules which coalesce to form scaly patches in which the hairs are broken because of hair shaft invasion. Under ultraviolet light (Wood's light) they show green fluorescence.

Diagnosis

Confirmation is by observing fungal elements in microscopic preparations or scraping of skin and nails. Species identification requires culture.

Treatment

Small local lesions are treated with topical imidazole (clotrimazole, miconazole, econazole). Extensive or chronic skin infections and all nail infections require oral therapy: griseofulvin or terbinafine for up to 6 weeks.

PITYRIASIS (TINEA) VERSICOLOR

This is caused by a yeast, *Malassezia furfur*, a commensal of human skin. In warm conditions the organisms flourish and change to mycelial forms producing hypo- or hyperpigmented macules with fine scales. Trunk or proximal limbs are common sites and lesions often coalesce. They fluoresce yellow/green under a Wood's light.

Diagnosis

The diagnosis can be confirmed by microscopy of scrapings after clearing with potassium hydroxide.

Treatment

A selenium sulphide shampoo or a topical imidazole cream is usually effective. In more florid cases, ketoconazole or itraconazole is preferable. Pigment changes may persist for months.

Anthrax

This is an infection commonly localised to skin, caused by *Bacillus anthracis*, a Gram-positive organism.

Epidemiology

- *Bacillus anthracis* causes septicaemic disease in wild and domestic animals in many areas of Asia and Africa. The bacilli form spores easily which can withstand drying and disinfection, survive in the animal carcass and contaminate the environment heavily
- Humans become infected through occupational exposure to infected animal products (wool, hide, bone, meat) and human anthrax is endemic in countries where animal anthrax is common
- Rarely, inhalation of heavily contaminated dust and ingestion of infected meat can cause primary anthrax pneumonia or gastroenteritis
- Anthrax is rare in Britain because of carefully supervised sterilisation and disinfection of imported animal products
- The IP is 2–7 days.

Clinical features

- Cutaneous anthrax usually starts as a small, itchy papule which grows in a few days into a thick-walled, off-white vesicle with a dark red or blackish base. Later the vesicle ruptures and a black crust (eschar) forms which lasts for several weeks. Moderate to severe surrounding oedema with smaller satellite pustules are common. Pain is rare
- Untreated cutaneous anthrax may spread to regional lymph glands and blood, causing septicaemia
- Pulmonary anthrax (severe respiratory distress), intestinal anthrax (abdominal pain) and septicaemic anthrax are highly fatal.

Diagnosis

This is by demonstrating anthrax bacilli in lesion material by Gram stain and culture. Serology is helpful. Orf is often confused with anthrax.

Treatment

Penicillin is the drug of choice, given IV for 7 days. In allergic patients, erythromycin, tetracycline or chloramphenicol can be used.

Prevention

This depends on controlling anthrax in livestock and by disinfecting imported animal products. Spores may rarely survive in bone meal fertilisers, and protective gloves should be worn by those handling it in bulk. Killed anthrax vaccine is recommended for workers exposed to imported hides and bones.

Lyme disease (Lyme borreliosis)

This is a zoonotic disease caused by a spirochaete, *Borrelia burgdorferri.*

Epidemiology

The vectors of Lyme borreliosis are ixodic ticks in whom the infection is maintained by horizontal transmission through different stages of their life cycle. The ticks are dependent on wild rodents and deer as feeding sources. Foci of infected ticks exist in the forests of the USA, Europe (including Britain), Russia, China, Japan and Australia. Transmission to man is through the bite of an infected tick, and does not take place until the tick has fed for several hours.

Clinical features

These are extremely variable but three overlapping stages are recognised.

Stage 1

Erythema chronicum migrans (ECM): 3–32 days after a tick bite a red papule develops at the site of the bite and gradually enlarges, with central clearing, to reach a size of 5 cm or more. More than one lesion may develop. Influenza-like symptoms may be present, sometimes with meningism. All symptoms may last for several weeks if untreated.

Stage 2

After weeks or months, and sometimes without a preceding ECM, the patient may develop symptoms pertaining to:

- nervous system: subacute meningitis, cranial nerve palsies, polyneuropathy
- cardiovascular: myocarditis, pericarditis, conduction defects
- musculoskeletal: myalgia, intermittent arthritis.

Stage 3

Months to years later, chronic neurological, skeletal or skin involvement may develop.

Diagnosis

This is based on serological tests (IgM and IgG antibodies) but is unreliable early and can remain negative if treated early. Skin biopsy yields the organism more easily than blood culture.

Treatment

Tetracycline for 10–30 days is usually effective for early disease (in children: penicillin or erythromycin). Later manifestations require longer courses. Ceftriaxone is useful for neurological disease.

Prevention

Important measures are: the avoidance of tick-infested areas; the wearing of protective clothing impregnated with diethyltoluamide; periodic search for unsuspected ticks on the body and their removal without leaving behind their mouth parts.

Gas gangrene (clostridial myonecrosis)

Gas gangrene is a potentially lethal condition, characterised by muscle necrosis and gas formation caused by a group of clostridial organisms which are anaerobic, spore-forming, Gram-positive bacilli.

Epidemiology and pathogenesis

The organisms usually involved are: *Clostridium perfringens*, *C. septicum*, *C.*

oedematiens and, rarely, *C. histolyticum*. Many are normal inhabitants of human and animal intestines. Infection results from wounds contaminated with soil or faeces. The organisms elaborate a wide range of toxins, of which alphatoxin is thought to be the most important.

In civilian practice, most cases are either following street or industrial accidents, or post-operative, particularly after lower-extremity or abdominal surgery.

Clinical features

- After an IP of 1–6 days, the area around the wound becomes swollen, painful, and the skin colour changes to magenta and then black. Bullae are common. Gas formation is indicated by crepitation (radiology confirms this). Exposed muscles look grey and the wound has an offensive, serosanguinous discharge
- The gangrene may extend rapidly and the patient becomes severely ill, with tachycardia, fever and hypotension. Renal failure and haemolytic anaemia are complications.

Diagnosis

- The presence of soft tissue gangrene, gas formation and evidence of muscle necrosis (visual, biopsy or computerised tomography (CT)) are pre-requisites for diagnosis. A Gram stain of the discharge showing clostridial organisms will be supportive but, on its own, may just indicate wound contamination or cellulitis
- Blood culture may be positive but is not significant in the absence of clinical features of gas gangrene.

Differential diagnoses

As well as clostridial myonecrosis, a number of other infections can give rise to necrosis with or without gas formation in and around a wound.

- Crepitant and anaerobic cellulitis: this can either be due to bacterioides, either alone or in combination with other facultative anaerobes such as *E. coli*, various streptococci and staphylococci
- Streptococcal necrotising fasciitis: *S. pyogenes* can cause an acute, rapidly progressive, sometimes fatal necrosis of skin, subcutaneous tissues and muscles surrounding a wound. Pain is usually late, there is prominent cutaneous erythema, with numerous pus cells in the wound discharge. The patient is seriously ill; and hypotension, DIC and respiratory, hepatic and renal failure are common. The fatality rate is high. Isolation of the organism from the affected site or from blood is necessary for confirmation
- Non-streptococcal necrotising fasciitis (synergistic necrotising cellulitis, Meleney's synergistic gangrene): skin, subcutaneous tissues and

fascial structures may be involved in a mixed infection of anaerobes and aerobes. This usually follows a traumatic wound and the abdomen, extremities, scrotum (Fournier's disease) and perineum are mostly involved. The prominent clinical findings are spreading erythema, oedema and tenderness. Bullae, subcutaneous gas formation and cutaneous gangrene appear. Muscle is not involved. The condition can be rapidly progressive and fatal.

Treatment

- The gangrenous area must be explored, with debridement of all soft tissues and muscles. This usually involves amputation in the case of limbs
- IV penicillin is the antibiotic of choice, but metronidazole is also effective and should be added, along with gentamicin if mixed infection is suspected
- Hyperbaric oxygen therapy may limit the spread of gangrene and allow conservation of partially viable tissues. A trial of such therapy should be given if local facilities exist before performing a more radical surgery
- Polyvalent anti-gas gangrene serum has doubtful value, may cause serious reactions, and is not recommended
- Necrotising fasciitis should be treated with a combination of antibiotics to cover possible aerobic and anaerobic infections (ceftotaxime, ampicillin and metronidazole) or penicillin IV if streptococcal, plus surgical debridement.

Prevention

- Thorough surgical cleanliness of traumatic wounds, removal of foreign bodies and the avoidance of ischaemia are the most effective preventive measures
- Patients undergoing above-the-knee amputation for arterial insufficiency are at risk and should receive pre-operative penicillin or metronidazole prophylaxis.

Leprosy

Epidemiology

Leprosy (Hansen's disease) is a chronic granulomatous disease caused by *Mycobacterium leprae*, affecting the skin, peripheral nerves and mucous membranes. The disease is endemic in Africa, the Indian subcontinent, South-East Asia, Central and South America, with millions of cases worldwide. Infection is spread by *M. leprae* in nasal discharges from lepromatous patients to the nose or through the skin of close contacts. The risk of transmission is small in other forms of leprosy. There is a high degree of natural resistance to the disease and 90 per cent of exposed people show no sign of infection.

Clinical features and diagnosis

The clinical manifestations of the disease depend on patients' immune response. The clinical and pathological states form four main classes:

Indeterminate leprosy is the earliest manifestation, and consists of a small, irregular, pale skin lesion, anywhere on the body. There is no sensory loss and the lesion usually heals spontaneously. At this stage, the pathology is a nonspecific inflammatory reaction. It may progress to any of the three later types, depending on the cell-mediated response

Tuberculoid leprosy is characterised by one, or a few, solitary sensationless macules with a raised edge or plaque. In dark skins, the lesions are depigmented (Plate 6a, between pp. 180 and 181). In pale skins, the lesions' colour is reddish. Affected nerves are thickened – these are commonly the ulnar, median, superficial radial, common peroneal and greater auricular nerves – with corresponding sensory and motor changes. The infection is met by a vigorous CMI response, which confines the disease to local lesions in the skin and nerves. The histology is of non-caseating epithelioid granulomas, without mycobacteria. This type of disease is probably not infectious

Lepromatous leprosy has numerous skin lesions, which are symmetrically distributed, and macular, plaque-form or nodular. There is no CMI response and the lesions are dominated by foamy histiocytes full of bacilli. The skin becomes thickened by this infiltration, and the ears, lips and nose swell (Plate 6b). The deepened creases on the face give a leonine appearance. Proliferation of the infection in the mucous membranes causes nasal congestion and keratitis. Increasing peripheral neuritis can be complicated by neuropathic ulcers on the limbs, with increasing disfigurement and disability

Borderline leprosy falls between tuberculoid and lepromatous, with a mixed clinical picture. Either clinical and immunological type can predominate or vary, as the state is unstable

Lepra reactions are acute exacerbations of the two polar types. Type 1 is due to increased cellular hypersensitivity in tuberculoid leprosy, and causes acute inflammation of the lesions. Type 2 reactions, in lepromatous leprosy, are due to vasculitis, producing erythema nodosum, iritis, orchitis and nephritis.

The diagnosis is usually established by the clinical picture and by biopsy of a skin lesion or thickened sensory nerve.

Treatment

Lepromatous and borderline leprosy are treated with a combination of rifampicin, clofazimine and dapsone for 2 years. The skin and neuropathic limb deformities may need surgical repairs. Tuberculoid leprosy is treated with rifampicin plus dapsone for 6 months. Type I lepra reactions respond to steroids, and thalidomide is useful for type 2 reactions.

Prevention

Bacille Calmette–Guérin vaccination gives some protection against leprosy and dapsone may be offered to child contacts of lepromatous cases.

CHAPTER 11

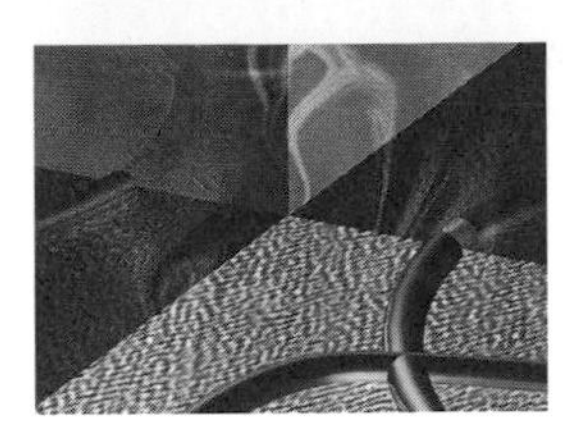

Infections of the Gastrointestinal Tract

CLINICAL SYNDROMES

Diarrhoea

Infective diarrhoeas (Table 11.1) are a major public health problem worldwide. In the developing countries they are a major cause of morbidity and mortality, particularly among children, and poor standards of sanitation and hygiene, undernourishment, overcrowding and inadequate medical resources are responsible factors. Infective diarrhoea is also common in the developed countries but mortality is very low.

Distinguishing features

Character of stools	*Site of pathology*	*Likely organism*
Large volume, watery. Pus and blood absent (naked eye/microscopy)	Small intestine (enteritis)	Viruses, *Vibrio cholerae*, *Escherichia coli* (EPEC, ETEC, EAggEC), toxin-type food poisoning, *Giardia*, cryptosporidium
Frequent, small volume. Pus and/or blood present (naked eye/microscopy)	Large intestine (colitis)	*Shigella*, *Entamoeba histolytica*, *E. coli* (EHEC, EIEC), *Clostridium difficile*

CAUSATIVE ORGANISMS OF DIARRHOEA

Bactenal	Viral	Parasitic
Campylobacter *Salmonella* *Shigella* *Escherichia coli* *Vibrio cholera* and other vibrios *Clostridium perfringens*, *C. difficile* *Yersinia enterocolitica* *Plesiomonas shigelloides* *Aeromonas hydrophilia* *Bacillus cereus*	Rotavirus Adenovirus Small round structured virus (SRSV) Calcivirus Astrovirus	*Giardia lamblia* *Entamoeba histolytica* *Cryptosporidium* *Cyclospora* *Microsporidia* *Isospora belli*

NB. In Britain, *Campylobacter*, *Salmonella* and *Shigella* are the most common bacterial causes of adult diarrhoea. In young children, most cases are due to viral infections, rotavirus accounting for the majority, followed by enteric adenoviruses

Table 11.1. Organisms which cause infective diarrhoea

Character of stools	*Site of pathology*	*Likely organism*
Initially large volume watery stools changing to small volume stools (macroscopic or microscopic blood/pus present)	Small and large intestines (enterocolitis)	*Salmonella*, *Campylobacter*, *Shigella sonnei*

Complications

- Dehydration, renal failure
- Hypernatraemic dehydration in children (particularly in obese): high serum sodium (>150 mmol/litre) despite overall body deficit of sodium, because of greater loss of water than sodium through diarrhoea and sometimes via skin. Concurrent hypertonicity of intracellular compartment may cause brain damage
- Septicaemia (*Salmonella*, *Yersinia*, *Campylobacter fetus*)
- Toxic colonic dilatation (*Salmonella*, *Campylobacter*, *Shigella*, *C. difficile*)
- Haemolytic–uraemia syndrome (enterohaemorrhagic *E. coli* O157, *Shigella dysenteriae*)

- Post-infective irritable bowel syndrome
- Secondary lactose intolerance
- Post-infective intestinal malabsorption (tropical sprue)
- Reactive arthritis (*Shigella*, *Salmonella*, *Campylobacter* particularly in HLA B29-positive individuals)
- Erythema nodosum (*Salmonella*, *Campylobacter*, *Yersinia enterocolitica*).

Investigations

- Stool culture
- Microscopy of faecal smear for pus cells and red cells (helps to identify the site of involvement, hence, the likely organism)
- Stool microscopy for ova, cyst and parasites (if history of foreign travel or persistent diarrhoea)
- Search for viruses (electron microscopy (EM), enzyme-linked immunoadsorbent assay (ELISA) for rotavirus, etc.)
- Blood culture (in severely ill patients)
- Urea and electrolytes
- Plain X-ray of abdomen (if distended abdomen)
- Sigmoidoscopy and rectal biopsy (if underlying inflammatory bowel disease suspected)
- Test for reducing substance in stool (if lactose intolerance suspected)
- Tests for intestinal malabsorption and bacterial overgrowth (if post-infective persistent diarrhoea syndrome suspected).

Differential diagnoses

Stool type	*Acute*	*Persistent*
Bloody	Intussusception Ischaemic colitis Inflammatory bowel disease Diverticulitis	Inflammatory bowel disease Colonic cancer
Non-bloody	Retrocaecal appendicitis Systemic infections in children	Lactose intolerance Milk intolerance Coeliac disease Overflow incontinence

Treatment

The management of infective diarrhoea consists primarily of the prevention and correction of dehydration.

Assessment of hydration

Level of hydration	*Condition of patient*
No dehydration	Patient well and alert Drinking normally (not thirsty) Urine volume normal Mouth and tongue moist Tears present Skin tone normal Fontanelle normal (in children)
Some dehydration (equivalent to 5–10% loss of body weight)	Patient restless or irritable Increased thirst Passing urine infrequently Mouth and tongue dry Eyes slightly sunken Tears absent Skin less elastic (skin tone may be normal in children with hypernatraemic dehydration; suspect this if the child is hyperventilating and has suffused conjunctivae) Fontanelle depressed (children)
Severe dehydration (>10% loss of body weight)	Patient apathetic, drowsy or floppy Drinks poorly or not at all Very sunken eyes Mouth and tongue very dry Passing urine scantily or not at all Skin grossly inelastic Fontanelle very depressed (children)

Rehydration therapy

Hydration status	*Action*
No dehydration, mild non-watery diarrhoea	Increase intake of fluids (water, tea, fruit juice, etc.)
No dehydration, profuse watery diarrhoea	Oral rehydration solution (ORS)*: Children <2 years: 50–100 ml after each stool

Hydration status	*Action*
	Children 2–10 years: 100–200 ml after each stool Adults and older children: as much as they want *plus*: the normal daily requirement of fluids and small amounts of light, calorie-rich foods†
Some dehydration	Calculated fluid deficit (= % dehydration × 10 × kg body weight in ml) should be given as ORS over 4–6 hours in small, frequent feeds. After rehydration, feeds are given hourly until day's requirement is complete (normal fluid requirement is 150 ml/kg in young children, plus continued loss through diarrhoea)†‡
Severe dehydration, or patient vomiting or unable to drink adequately	IV hydration using, initially, plasma or 0.9% NaCl to correct shock, then 0.18% NaCl in dextrose to correct deficit. Resume ORS when feasible‡

* The presence of an optimum level of glucose in ORS actively promotes absorption of Na^{+} and Cl^{-} (and water) through intestinal mucosa, despite diarrhoea. In Britain, ORS solution normally used contains less sodium than the WHO/UNICEF-recommended solution:

Constituents	*Dioralyte sachet (mmol/litre)*	*WHO/UNICEF ORS – citrate (mmol/litre)*
Sodium	60	90
Potassium	20	20
Chloride	60	80
Citrate	10	10
Glucose	90	111

† In Britain, it is customary to withhold milk feeds in the treatment of childhood diarrhoea, and to give ORS alone for 12–24 hours. However, continued feeding as recommended by the WHO is safe and should certainly be practised in developing countries to prevent undernutrition from recurrent episodes of diarrhoea.

‡ In the presence of hypernatraemia, it is important to correct fluid deficit slowly over a period of 24–48 hours. Too rapid lowering of serum sodium may precipitate a rapid shift of water in the hypertonic intracellular compartment, causing cerebral oedema and convulsion.

Refeeding

The usual milk feeds can generally be introduced after 12–24 hours, even if diarrhoea continues. Regrading of feeds is unnecessary. Older children and adults should start taking a light, calorie-rich diet. ORS drinks should continue to be given in the presence of diarrhoea.

Test for lactose intolerance (check stools for presence of reducing substances >0.5 per cent) if introduction of cow's milk feed worsens diarrhoea. If positive, a low-lactose soya-based milk should be used.

Specific antimicrobial therapy

This is unnecessary for most cases of acute diarrhoea, which are self-limiting and short-lasting. Antibiotics are indicated in:

- severe *Shigella*, *Salmonella* and *Campylobacter* infections
- cholera
- severe *Y. enterocolitica* infection
- *Clostridium difficile* colitis
- giardiasis
- amoebic dysentery
- traveller's diarrhoea.

Other measures

Anti-motility drugs such as diphenoxylate and loperamide have little role in the treatment of infective diarrhoea.

Prevention

- In hospitals, all patients with diarrhoea should be in isolation with safe handling of faeces and other contaminated materials
- All cases should be excluded from work or school or day nurseries until symptom free, except for cases of salmonellosis where food handlers, whose work involves touching unwrapped foods to be consumed raw or without further cooking, should have three consecutive negative faecal cultures at weekly intervals before returning to work. Prolonged exclusion of young children below 5 years of age from day nurseries, play groups or nursery schools is not necessary if supervision of hygiene is adequate. Meticulous attention to standards of personal hygiene, hand-washing facilities and supervision of hygiene in young children are necessary for the prevention and control of institutional outbreaks of shigellosis and giardiasis
- Public water supplies should be protected from human and animal faeces and should be filtered properly if such risks exist. Travellers should only drink boiled or bottle waters and avoid unfiltered lake or

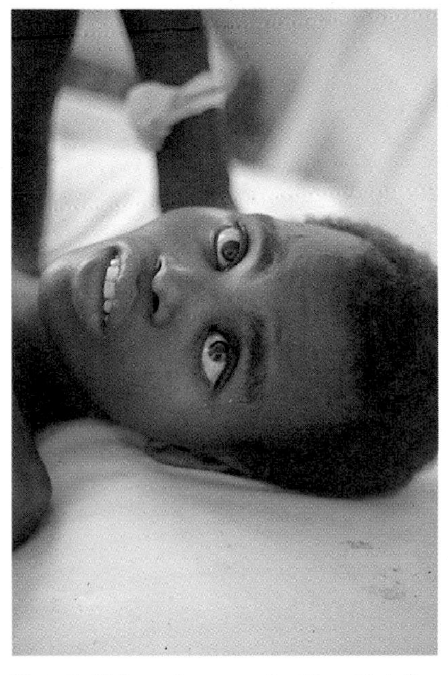

Plate 1 Tuberculous meningitis with left sixth nerve palsy.

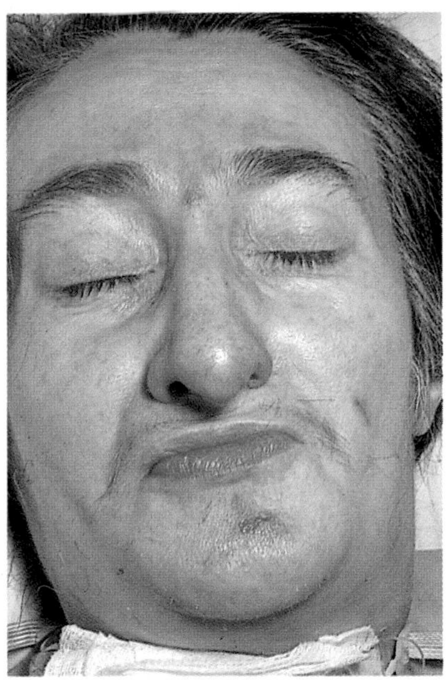

Plate 2 Risus sardonicus in tetanus.

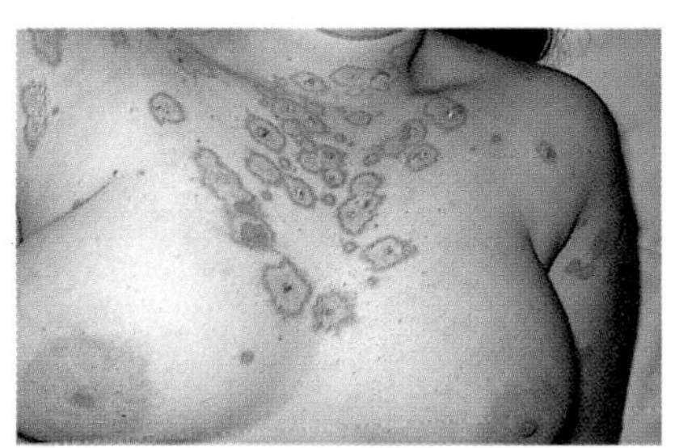

Plate 3 Erythema multiforme with characteristic target lesions.

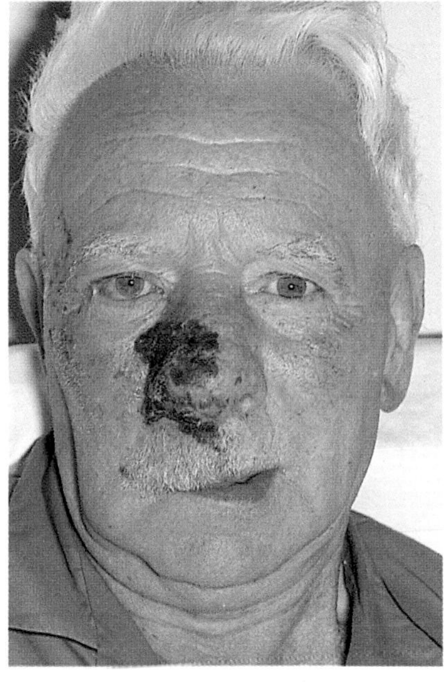

Plate 4 (right) Seventh nerve palsy complicating maxillary herpes zoster (Ramsey Hunt syndrome).

(a) 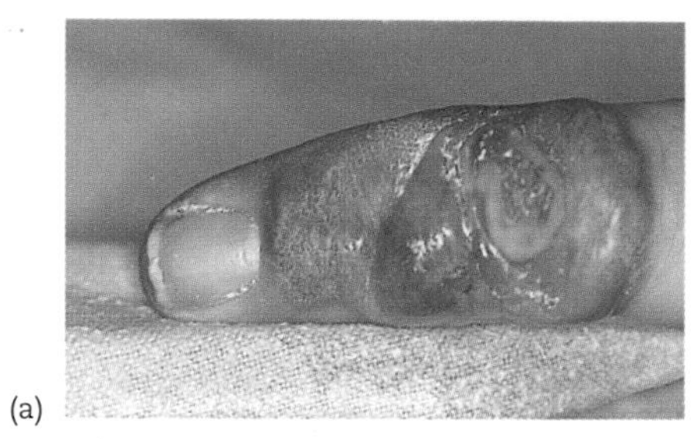 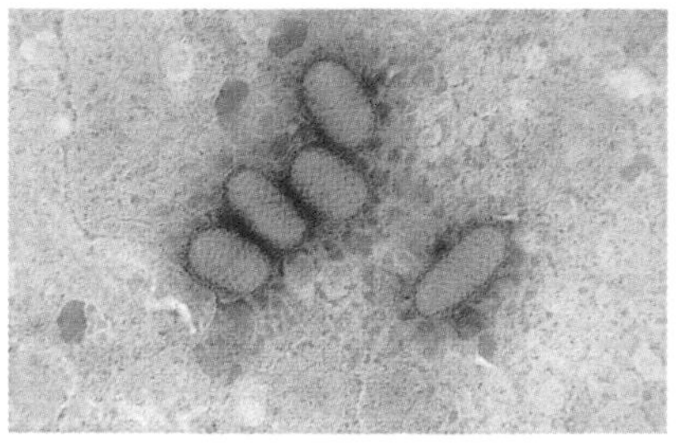 (b)

Plate 5 Orf in a sheep farmer (a) with the electron micrographic appearance of the virus (b).

(a) 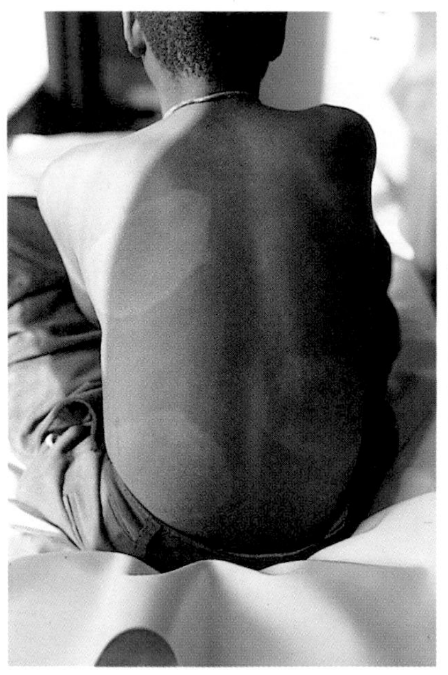 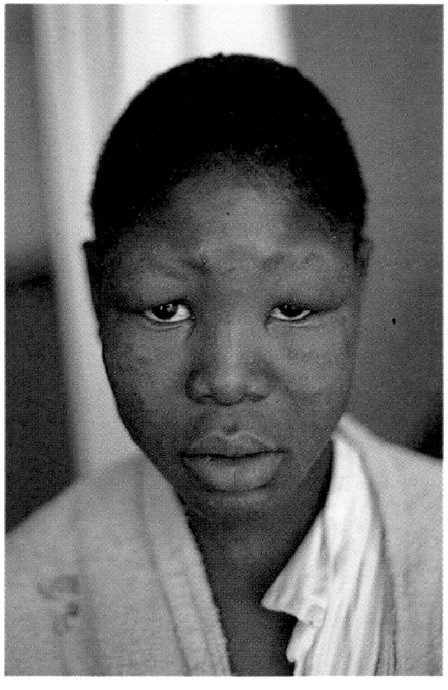 (b)

Plate 6 Tuberculoid (a) and lepromatous (b) leprosy.

Plate 7 Intravascular haemolysis complicating falciparum malaria on admission (left) and after 4 days quinine therapy (right).

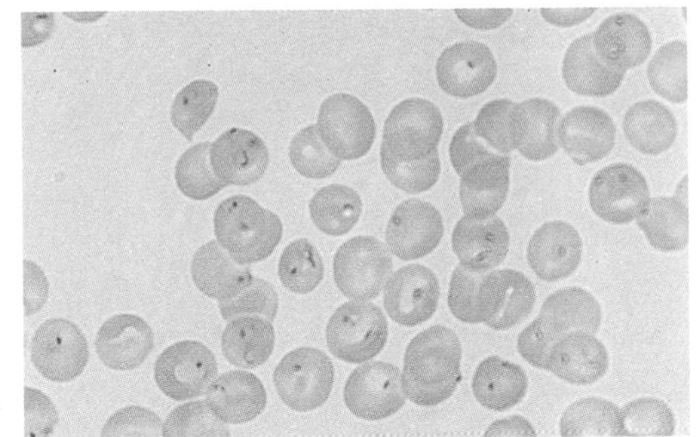

Plate 8 *Plasmodium falciparum* heavy parasitaemia.

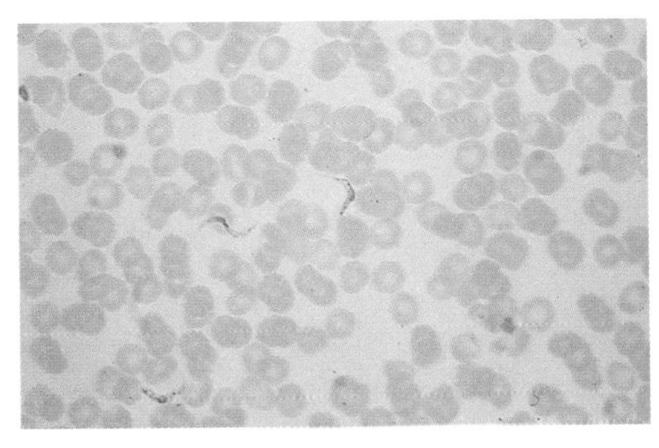

Plate 9 East African trypanosomiasis contracted on safari.

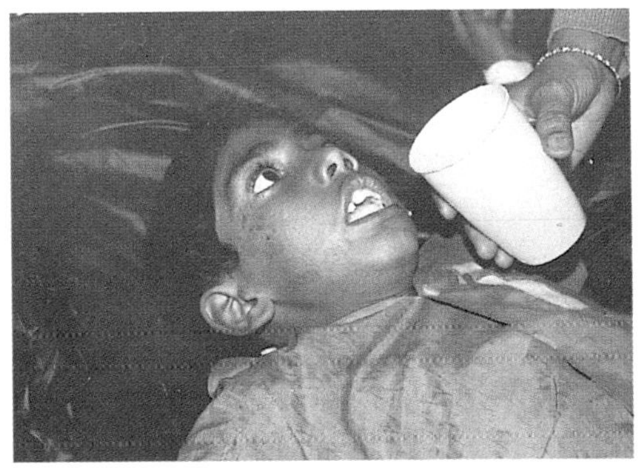

Plate 10 Hydrophobia in rabies.

stream water, particularly in mountainous areas. Water purifying tablets are unreliable against parasitic diarrhoea

• Food of animal origin should be thoroughly cooked. Eggs should be boiled for at least 3 min. High standard of hygiene must be observed in all areas where raw meat is handled, i.e. slaughterhouses, meat shops, restaurants and domestic kitchens. Raw and cooked food should be kept separate. Frozen food needs complete thawing before cooking

• In developing countries, provision of safe drinking water, sanitary disposal of excreta and public education on food hygiene are important measures.

SPECIFIC INFECTIONS

Helicobacter pylori infection

Epidemiology and clinical features

Current evidence suggests that these Gram-negative curved micro-aerophilic organisms are closely involved in the aetiology of the following gastro-duodenal disorders:

- short-duration dyspepsia associated with acute gastritis during initial colonisation. This may be self-limiting
- most cases of type B chronic gastritis. This may be associated with chronic dyspesia or be asymptomatic
- chronic duodenitis
- peptic ulcer diseases, particularly duodenal ulcers
- possibly gastric cancers.

The source and mode of transmission remain unknown.

Diagnosis

Confirmation is usually by gastroscopy (urease test, histology and culture of biopsy specimen). Urease breath test and serology are also helpful.

Treatment

Helicobacter pylori can usually be eradicated by administering a 2-week course of omeprazole (20 mg b.d.), metronidazole (400 mg three times daily) and tetracycline (500 mg four times daily). Treatment is useful in symptomatic cases of *H. pylori*-associated chronic gastritis and in relapsed duodenal ulcer.

Shigella infection (bacillary dysentery)

Epidemiology

- These Gram-negative organisms belonging to the genus *Shigella* are an important cause of acute diarrhoea throughout the world. There are four subgroups (or species) of these strictly human pathogens, based on biochemical characteristics and antigenic differences: subgroup A (*S. dysenteriae*) subgroup B (*S. flexneri*), subgroup C (*S. boydii*) and subgroup D (*S. sonnei*)
- *Shigella flexneri* and *S. dysenteriae* produce more serious illnesses and occur mainly in the developing countries, whilst *S. sonnei* is endemic in the developed countries
- Outbreaks are common in camps, prisons, day-care centres and infant schools. Transmission is faecal–oral, either from person to person, or water- and food-borne. Flies may spread infection in the tropics
- Children are the most commonly affected in the UK. *Shigella sonnei* accounts for the majority of cases. Most *S. flexneri* infections have their origins abroad
- The IP is 2–4 days.

Pathogenesis

Shigella organisms invade the colonic mucosa causing inflammation, ulceration, haemorrhage and sloughing, and cause the characteristic bloody mucoid diarrhoea of *Shigella* dysentery. There is also an associated enteropathy early in the illness, resulting in secretion of isotonic fluid into the lumen of the jejunum. This is probably enterotoxin induced.

Shigella dysenteriae type I also elaborates an entertoxin (Shiga toxin) which has neurotoxic, cytotoxic and enterotoxic properties in experimental animals. It is linked with the production of the haemolytic–uraemic syndrome which sometimes complicates *S. dysenteriae* infection.

Clinical features

- The illness begins abruptly with headache, fever and abdominal discomfort. Diarrhoea follows and initially it may be watery and without any visible blood (reflecting the initial enteropathy)
- In many *S. sonnei* infections, the diarrhoea may remain watery and settles within 3–5 days, but in others and in *S. flexneri* and *S. dysenteriae* infections blood and mucus appear in stools within a day or two
- In severe cases the classic picture of dysentery appears, with severe abdominal cramps, tenesmus and passage of very small volume stools consisting just of blood, pus and mucus

• In young children hyperpyrexia is not uncommon and convulsion may occur, sometimes at the onset of the illness
• Convalescent carriage after clinical recovery is common but usually ceases by 8 weeks.

Complications

• Rare in average cases of *S. sonnei* infection
• Toxic dilatation and/or perforation (*S. flexneri* and *S. dysenteriae*)
• Haemolytic–uraemic syndrome (*S. dysenteriae*).

Diagnosis

Although the clinical diagnosis may be suspected when typical dysenteric features are present, definitive diagnosis depends on isolation of the organism by faecal culture. Polymorphonuclear leucocytosis is common.

Treatment

• In the UK, most cases (usually due to *S. sonnei*) are mild and of short duration, and require only attention to adequate oral fluid intake
• In severe cases, which are usually due to *S. flexneri* and, rarely, to *S. dysenteriae*, antibiotics are indicated. In children the drug of choice is nalidixic acid; in adults, ciprofloxacin or norfloxacin.

Prevention

See p. 180.

Campylobacter infection

Epidemiology

• The curved, motile, Gram-negative *Campylobacter* bacilli are a major cause of diarrhoea in humans. Most human infections are due to *C. jejuni*, whereas *C. fetus* generally behave as an opportunistic pathogen, producing invasive illness in debilitated or immunosuppressed patients with metastatic localisations in extra-intestinal sites
• In Britain, *C. jejuni* is the most common bacterial cause of acute diarrhoea
• Children and young adults are most commonly affected. Asymptomatic infections are more common in the tropics
• The organisms infect the intestines of many domestic and wild animal species, which are the ultimate source of all human infections. Among food animals, poultry is particularly important. Transmission is through contaminated food, milk or water, or through contact with infected pets

- Direct person-to-person transmission is rare
- The IP is 3–5 days.

Pathogenesis

Campylobacter jejuni is an invasive organism which produces inflammatory changes in the intestinal mucosa, particularly in the ileum and colon. An enteropathy also occurs and accounts for initial watery diarrhoea.

Bacteraemia is rare with *C. jejuni* infection.

Clinical features

- Fever, myalgia and abdominal pain may occur for up to 2 days before onset of diarrhoea. The stools are large-volume, watery and offensive, but later may become small-volume, bloody and mucoid
- Vomiting is common in children. Abdominal pain is very common, is usually colicky in nature, and is relieved by defaecation
- The diarrhoea ceases within 1 week, but low-grade abdominal pains may continue for several more days with some degree of intestinal hurry
- Most patients cease to excrete the organism in their faeces within 3–4 weeks.

Complications

- Acute ileitis with pain and tenderness in the right lower abdomen, which may mimic acute appendicitis
- Acute colitis with prominent bloody diarrhoea and sigmoidoscopic evidence of acute colitis. Biopsy generally shows histological features typical of infective colitis but, occasionally, changes suggestive of inflammatory bowel disease (IBD) may be present
- Guillain–Barré syndrome (GBS)
- Rarely: acute cholecystitis, erythema nodosum and reactive arthritis.

Diagnosis

This is confirmed by positive faecal culture.

Treatment

Patients are usually recovering by the time of diagnosis, and antibiotics are only indicated in patients with severe or prolonged diarrhoea and in the rare bacteraemic cases. The drugs of choice are either erythromycin or ciprofloxacin for 5 days. Faecal cultures become rapidly negative following such therapy.

Prevention

See p. 180.

Salmonellosis

Epidemiology

• Salmonellosis is caused by the non-typhoid salmonellae which are primarily parasites of animals, inhabiting their intestines. Of about 2000 serotypes, only a small number account for the vast majority of human infections. Examples of some of the commonly isolated serotypes from human cases are *S. enteritidis*, (most common in the West) *S. typhimurium*, *S. virchow*, *S. hadar*, *S. heidelberg*, *S. agona* and *S. indiana*

• Salmonellosis occurs more commonly in the developed countries. This is primarily due to the increased dependence on foods of animal origin, which has led to large-scale intensive farming methods for food animals creating conditions suitable for the rapid spread of infection among the animals

• Transmission is through ingestion of food or drink contaminated with *Salmonella* from an animal source. Inadequately cooked poultry, beef, pork, hen's eggs (infected via the oviduct) and raw milk (in some countries) are common examples

• Person-to-person transmission is infrequent. Most institutional outbreaks are food related but may follow admission of patients with undiagnosed *Salmonella* diarrhoea, particularly in geriatric or maternity/neonatal units. Turtles and other pets are occasional sources

• The IP is normally 12–48 hours.

Pathogenesis

Two mechanisms are responsible for the diarrhoea: enterotoxin-induced fluid and electrolyte transport defect in the small intestine, and invasive mucosal inflammation in the colon and lower ileum.

Clinical features

• The illness begins abruptly with nausea and vomiting, often associated with malaise, headache and fever, followed by cramp-like abdominal pains and diarrhoea. Initially, the stools are large-volume and watery without blood (toxin-induced enteropathy phase), but later often become small-volume and bloody (colitis phase)

• The severity of diarrhoea is variable, from a mild attack lasting no more than a few hours to frequent voluminous watery stools. The average cases settle within a few days and persistence of diarrhoea beyond 3 weeks is rare

• After recovery, patients usually excrete *Salmonella* for 4–6 weeks, but longer in extremes of ages. Chronic carriage beyond 1 year is rare (less than 1 per cent).

Complications

Severe dehydration and renal failure: these are prone to occur in the elderly, the immunocompromised and in persons with gastric achlorhydria from any cause

Colitis: with severe bloody diarrhoea. Rarely, toxic megacolon may develop (Fig. 11.1). On sigmoidoscopy the mucosa is diffusely inflamed with or without ulceration and biopsy will show typical changes of infective colitis and sometimes of IBD

Ileitis: with pain and tenderness localising over the right iliac fossa. This may be confused with appendicitis

Invasive salmonellosis: this may present either as septicaemia or as a typhoid-like illness without significant diarrhoea, or as a metastatic infection in extra-intestinal sites, e.g. meninges, bones and joints, lungs, spleen, kidneys, heart valves and atheromatous blood vessels. Host

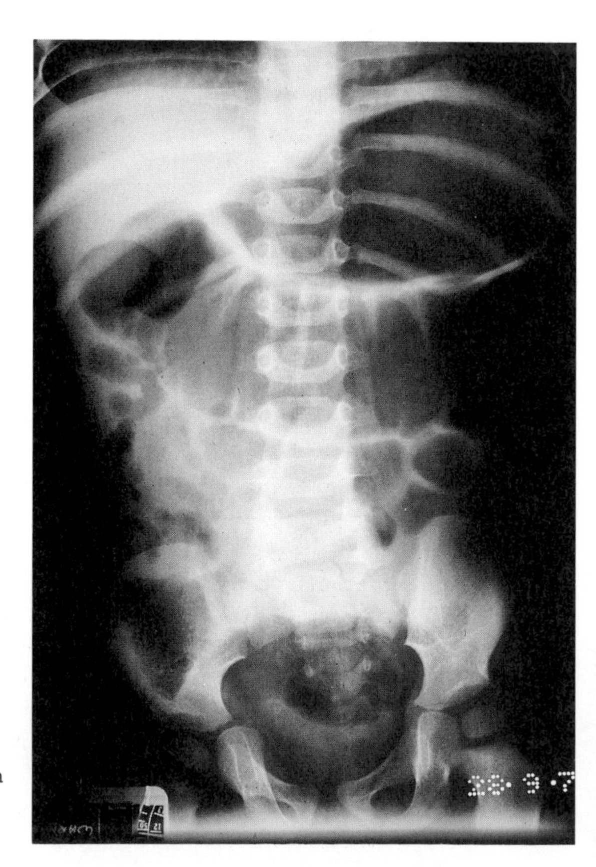

Fig. 11.1. Toxic megacolon complicating *Salmonella* colitis.

factors (extremes of age, immunosuppression, sickle cell disease, debility, gastric achlorhydria) and certain serotypes (*S. cholerae-suis*, *S. virchow*, *S. dublin*, multiresistant *S. typhimurium* with virulence plasmids in the tropics) are important determinants of invasive disease

Post-infective irritable bowel syndrome

Reactive arthritis: affecting one or more joints (commonly knee and ankles) during the 2nd or 3rd week of illness, particularly in individuals with HLA-B27 antigen.

Diagnosis

Definitive diagnosis depends on isolation of the organisms by stool culture, as *Campylobacter* and *Shigella* infections cause similar symptoms. Blood culture should be done in all severely ill patients.

Treatment

- Most cases have a short-lasting, self-limiting illness which requires only attention to oral rehydration and does not require antibiotic therapy
- Patients with invasive illness, and those who are prone to develop invasive illness (see above) or have severe symptoms, should receive antibiotics (ciprofloxacin in adults, cefotaxime or ceftriaxone in children).

Prevention

See p. 180.

Diarrhoea due to *Escherichia coli*

Five major categories of diarrhoea-producing *E. coli* are currently recognised: enteropathogenic *E. coli* (EPEC), enterotoxigenic *E. coli* (ETEC), enteroinvasive *E. coli* (EIEC), enterohaemorrhagic *E. coli* (EHEC) and enteroaggregative *E. coli* (EAggEC).

Enteropathogenic *E. coli*

- These are identified by virtue of their possession of specific O antigens which are agglutinable by specific antisera. The important serogroups are O18, O26, O44, O55, O86, O111, O114, O119, O125, O126, O128 and O142. Their virulence is probably related to the possession of plasmid-mediated adherence factor. Diarrhoea results from the dissolution of intestinal microvilli
- They are an important cause of infantile gastroenteritis in developing countries, but have become rare in developed countries

• Transmission is faecal–oral, often through contamination of infant feeds. Faulty hand washing of the attenders has been responsible in many institutional outbreaks
• Acute-onset watery diarrhoea without macroscopic or microscopic blood and pus is the characteristic finding, and fever is usually absent. The diarrhoea is often short-lasting and self-limiting but, in epidemics involving children's hospitals and nurseries, the clinical course is often more severe and protracted, the infectivity and mortality higher
• Oral rehydration remains the mainstay of treatment but in severe, protracted diarrhoea, co-trimoxazole may be effective. In institutions, infants should be nursed under strict enteric precautions.

Enterotoxigenic *E. coli*

These strains produce heat-labile (LT) and heat-stable (ST) toxins which produce diarrhoea through stimulation of the adenylate cyclase/cyclic adenosine monophosphate (AMP) or guanylate cyclase/cyclic guanosine monophosphate (GMP) pathways. They are an important cause of diarrhoea in children and adults in developing countries and also commonly affect travellers from the West to these countries. Profuse watery diarrhoea and abdominal cramp are common features. Transmission is via contaminated food or water. Ciprofloxacin shortens the duration of diarrhoea.

Enteroinvasive *E. coli*

These strains can cause a dysentery-like illness by producing invasive inflammatory colitis. The infections are not uncommon in the tropics but rare in the developed countries. Ciprofloxacin should be used in severe cases.

Enterohaemorrhagic *E. coli*

• These strains are characterised by their ability to produce Shiga-like toxin (also known as verotoxin) and they possess a plasmid-mediated fimbria enabling their attachment to intestinal mucosa. The most commonly implicated serotype in human illness is O157:H7, but other serotypes are occasionally involved
• EHEC strains are an important cause of bloody diarrhoea in Europe and North America with increasing incidence of both sporadic cases and outbreaks. The source of infection is often difficult to establish in sporadic cases, but undercooked processed ground beef is an important source. Person-to-person transmission, contaminated water and unpasteurised milk have also been responsible. *Escherichia coli* O157:H7 has

been found in beef, pork, lamb and poultry samples. In the USA, cattle appear to be an important source

- Typical clinical features are cramp-like abdominal pains of sudden onset, followed by diarrhoea, often bloody. It is unusual to find pus cells in faecal smears and fever is usually absent or low-grade. Diarrhoea usually settled within 8 days. The role of antibiotics is uncertain
- EHEC infection may be complicated by haemolytic–uraemic syndrome. Diagnosis is by demonstrating *E. coli* O157 or verotoxin in faecal specimens (consult laboratory). Serology may help.

Enteroaggregative *E. coli*

These strains have been recently recognised as a cause of infantile gastroenteritis (both acute and persisting) in some of the developing countries. Diagnosis requires tissue culture assay, to demonstrate the characteristic aggregation pattern among the organisms and their attachment to cells.

Cholera and other vibrios

Epidemiology and pathogenesis

Cholera is an acute diarrhoeal illness caused by organisms belonging to the species *V. cholerae*. Of the 139 O serogroups, only the serogroup O1 (classical and eltor biotypes) and O139 are known to cause epidemic cholera. An identical enterotoxin (cholera toxin) is elaborated by these organisms, which activates the enzyme adenylate cyclase causing accumulation of cyclic AMP in the intestinal mucosa and an outpouring of isotonic fluid into the lumen, resulting in the characteristic watery diarrhoea.

- Seven cholera pandemics have occurred during the past two centuries, all originating from the Ganges delta in the Indian subcontinent. Classic cholera biotype accounted for the initial six, but during the 1960s the eltor biotype arrived in the delta from Indonesia, replacing classic cholera, and began spreading out to other countries of South and South-East Asia, to the Middle East, Africa and parts of Europe, causing the seventh pandemic, and in 1991 spread to South America. However, since 1992 a previously unknown serogroup, O139 (*V. cholera* Bengal), has been causing epidemics of severe cholera-like illness in the Indian subcontinent and other Far Eastern countries
- *Vibrio cholera* exists widely in natural aquatic environments in association with zooplankton, and the existence of environmental reservoirs of

serogroup O1 in the Gulf coast of the USA and in Queensland, Australia, has been recognised

- Humans are the only natural hosts of cholera-producing organisms
- Transmission is usually through drinking water contaminated with the faeces or vomitus of an infected person, and less commonly via food. Raw or undercooked seafood harvested in polluted waters have caused outbreaks.

Clinical features

The IP is usually 2–3 days. The stool is white to yellow with flecks of mucus ('rice-water'), and has a fishy odour. The illness presents acutely with rapid onset of vomiting and profuse watery diarrhoea, accompanied by abdominal colic and leg cramps. Fever is absent. The rapid dehydration may lead to hypovolaemic shock and death within hours of onset. Mild and asymptomatic infections are not uncommon. Convalescent faecal carriage often lasts for several weeks.

Diagnosis

Rapid presumptive diagnosis can be obtained by dark field microscopy showing motile vibrios. Stool culture is necessary for distinguishing cholera from other causes of watery diarrhoea.

Treatment

- The mainstay of treatment is early replacement of fluid and electrolyte loss (see p. 177)
- Ciprofloxacin or norfloxacin are effective in shortening the duration of the clinical illness, eliminating the *Vibrio* from faeces. Close contacts should receive chemoprophylaxis. In an outbreak situation, the source of infection should be investigated
- Parenteral, killed whole-cell cholera vaccines give poor, short-lasting protection and are of little practical value.

DISEASE CAUSED BY OTHER VIBRIOS

- *Vibrio cholera* non-O1: serogroups of *V. cholera* other than O1 and O139 are known to cause sporadic cases of acute diarrhoea in many countries. These organisms are commonly found in environmental waters and transmission is through consumption of raw or undercooked seafood
- *Vibrio parahaemolyticus:* these organisms are widely distributed in coastal waters. They are a common cause of both sporadic cases and outbreaks of diarrhoea in many countries, but particularly in Japan and other Far East/South-East Asian countries. Raw or undercooked

seafoods are the main vehicle of transmission. Acute watery diarrhoea with abdominal cramps is the most common manifestation but bloody stools and high fever may also occur.

Colitis due to *Clostridium difficile*

Clostridium difficile is an important cause of colitis in people who have received antibiotics. In severe cases there are characteristic inflammatory changes in the colonic mucosa (pseudomembranous colitis (PMC)).

Epidemiology and pathogenesis

- The organisms are large, Gram-positive anaerobic bacilli with terminal spores. They are found in stools of about 3 per cent of healthy adults and in a higher proportion of hospitalised adults, particularly those receiving antibiotics
- Two types of toxin, an enterotoxin and a cytotoxin, are produced. The factors that cause *C. difficile* to proliferate in the adult gut, elaborate toxins and produce inflammatory changes are not clearly understood. It is likely that prior antibiotic therapy creates the necessary conditions in the bowel for *C. difficile*, present in the environment, to colonise and flourish
- The organisms can be transmitted within a hospital unit, through contaminated hands of staff, or by direct contact between patients or through equipment, e.g. sigmoidoscopes.
- Outbreaks are common. Sporadic cases in the community also occur
- Clindamycin and ampicillin are the antibiotics most commonly incriminated but association has been reported with most antimicrobial drugs
- Although *C. difficile* colitis can occur at any age, the elderly are involved most commonly, particularly those who are debilitated and have had abdominal surgery.

Clinical features

- Diarrhoea commonly begins 4–10 days after initiation of antibiotic therapy. The stools are usually watery but may contain blood. White mucoid material is commonly present
- Fever, abdominal pain and tenderness, and leucocytosis are associated features.

Complications

Hypoalbuminaemia, electrolyte disturbances and, rarely, toxic megacolon or perforation. Fatality is not uncommon in debilitated patients.

Diagnosis

- Demonstration of *C. difficile* cytotoxin in stool specimen
- Sigmoidoscopic changes vary from mild inflammation to multiple discrete, yellow-white plaques, which may become confluent (pseudomembrane)
- Rectal biopsy in PMC shows glandular dilatation, epithelial loss and outpouring of mucus, fibrin and neutrophils on to the surface.

Treatment

- In hospital, patients should be nursed under enteric precautions
- Mild cases may only need discontinuation of the offending antibiotic. In others, oral vancomycin or metronidazole are effective
- Relapses may occur following either treatment regimens
- Antimotility drugs should be avoided as they can precipitate toxic megacolon.

Rarer bacterial causes of diarrhoea

YERSINIA ENTEROCOLITICA INFECTION

Epidemiology and pathogenesis

This organism is an uncommon cause of acute diarrhoea. It belongs to the genus *Yersinia*, which also includes *Y. pestis*, the cause of plague, and *Y. pseudotuberculosis* which causes mesenteric adenitis, sepsis and reactive arthritis.

Asymptomatic carriage is common in animals, and pork is an important source of infection for humans. Water-borne and milk-borne infections occur. Person-to-person transmission has also been reported.

Clinical features

Acute febrile watery diarrhoea is normally the sole manifestation in young children. In older children and adults, right lower quadrant abdominal pain may be present as well, mimicking appendicitis (due to underlying ileitis or mesenteric adenitis). Spontaneous recovery usually occurs within a few days.

Complications

Septicaemia (in patients with iron overload, or the immunocompromised), reactive arthritis (particularly in individuals with HLA B27) and erythema nodosum.

Diagnosis and treatment

Isolation of the organism from faeces or by serology. Most cases are mild and self-limiting but severe cases should be treated with co-trimoxazole or ciprofloxacin.

PLESIOMONAS SHIGELLOIDES AND AEROMONAS HYDROPHILA INFECTIONS

These are aquatic micro-organisms found commonly in fresh and salt water, and human illnesses have followed consumption of contaminated shellfish and oysters. Most reports have come from East Asia and Australia (*A. hydrophila*).

Viral gastroenteritis

ROTAVIRUSES

- Of the five groups of rotaviruses, group A is the most common in humans. Group B has caused large outbreaks of adult gastroenteritis in China but is rare elsewhere. Group C rarely causes human infections but is common in animals, along with group B and D. The proximal small intestine is the site of infection and there is disorganisation of enterocytes and their functions
- Rotaviruses are the most important cause of childhood diarrhoea throughout the world. Although common in children below the age of 2 years, they are not infrequent in children up to the age of 4. Infections in adults occur but tend to be asymptomatic or mild until old age, with outbreaks occurring in nursing homes. There may be high asymptomatic carriage in neonates
- In temperate climates, almost all cases occur in winter. In warmer countries cases occur throughout the year
- Transmission is usually faecal–oral but may be respiratory. Affected persons excrete virus in their stools for about 1 week from the onset of illness
- After an IP of 1–3 days, illness begins with fever, vomiting and frequent watery diarrhoea of varying severity. Symptoms usually persist for 4–6 days. There is no specific therapy; attention to fluid intake is essential
- The diagnosis is by demonstrating rotavirus in stool specimens using EM or ELISA.

ENTERIC ADENOVIRUSES

Up to 10 per cent of community-acquired diarrhoea in young children is caused by these viruses. Unlike the adenoviruses which cause respiratory infections, the enteric adenoviruses are difficult to propagate in tissue cultures but identifiable in stools by EM. Types 40 and 41 are mostly involved. The clinical picture is indistinguishable from other types of viral gastroenteritis.

ASTROVIRUS

This is another virus capable of causing gastroenteritis in young children. Seroepidemiological studies suggest that astrovirus infections occur commonly worldwide.

SMALL ROUND STRUCTURED VIRUSES (SRSV) (OFTEN CALLED NORWALK AND NORWALK-LIKE VIRUSES) AND CALICIVIRUSES

- These viruses probably account for one-third of the non-bacterial gastroenteritis outbreaks, particularly in older children and adults
- SRSV are probably atypical caliciviruses
- The outbreaks typically occur in holiday camps or cruise ships, within families, in schools, hospitals and nursing homes, and following swimming in polluted water. Consumption of raw or poorly cooked shellfish is responsible for some outbreaks. The faecal–oral route is the probable method of transmission but ingestion of airbourne particles may also be important
- After an IP of 24–48 hours, there is an abrupt onset of nausea and vomiting. Diarrhoea may be absent, mild or severe. There may be headache, giddiness, abdominal pain and slight sore throat. Pyrexia, if present, is slight. Symptoms usually subside within 48 hours
- Diagnosis can be established by immune EM of stool and vomit to detect virus. Stools collected within 24 hours of onset are the best specimens.

Parasitic causes of diarrhoea

GIARDIASIS

Epidemiology

- Diarrhoea is due to infection by a flagellated protozoa, *G. lamblia* (*intestinalis*) and occurs worldwide
- The usual source of infection is another infected person, although animals such as beavers and other wild and domestic animals may also act as a source

- Transmission is faecal–oral, either from person to person, especially in institutions and day-care centres, or through contaminated drinking water
- Giardia may survive routine chlorination, and consumption of inadequately treated or unfiltered water is a frequent cause of infection in travellers to many tropical as well as some temperate climates
- The IP is about 2 weeks.

Pathogenesis

There is colonisation of the upper small intestine by the trophozoites which develop after ingestion of the cysts. Mucosal injury is characterised by partial or subtotal villous atrophy.

Clinical features

Many infections are asymptomatic. In others, there is watery diarrhoea of abrupt onset. This may settle within a few days but, in a significant number of cases, may persist in a less severe form for weeks or months. Intestinal malabsorption with steatorrhoea, abdominal cramps, flatulence and weight loss are common features.

Infected individuals often excrete cysts for weeks, or months if untreated.

Diagnosis

Examination of three freshly voided faecal specimens on separate days will reveal cysts or trophozoites in up to 95 per cent of cases. The diagnostic accuracy is improved by examining for trophozoites in duodenal fluid (via aspiration) or in duodenal biopsy.

Treatment

Metronidazole (for 3 days) or tinidazole (single dose) are effective in more than 90 per cent of cases. Relapses may occur. Mepacrine is also effective, but side effects are common.

Prevention

See p. 180.

CRYPTOSPORIDIOSIS

Epidemiology and pathogenesis

This is an important cause of infective diarrhoea in humans, caused by a coccidian protozoan, *Cryptosporidium* spp. The organisms are widely prevalent in the animal kingdom, including small and large domestic animals as well as poultry, birds, fish and reptiles, and are present as oocysts in their faeces. After ingestion by another host, the oocysts infect the small intestinal mucosa and multiply asexually by schizogony, then sexually into oocysts, which are excreted with the faeces. The oocysts can survive in the environment for long periods.

Human infections occur worldwide. Transmission is either direct faecal–oral (person-to-person, animal-to-person) or water-borne. Faecally contaminated natural water is an important source of infection for travellers and outbreaks have occurred worldwide from public water supplies. The organisms can survive routine chlorination. The IP is about 7 days.

Clinical features

- There is acute-onset watery diarrhoea and abdominal discomfort. Fever and vomiting may occur. In previously healthy immunocompetent persons, diarrhoea usually subsides within 14 days
- In immunocompromised individuals, especially patients with acquired immunodeficiency syndrome (AIDS), cryptosporidiosis is a serious illness with persistent profuse watery diarrhoea, lasting for weeks and sometimes months. Dehydration and malnutrition occur commonly, leading to death. In such patients, colonisation may involve the whole length of the intestine, from stomach to rectum. Extension into the biliary tract may cause cholangitis
- Infection can be asymptomatic.

Diagnosis

This is by demonstrating acid-fast oocysts in the faecal smear, using either Ziehl–Nielson (ZN) or safranin–methylene blue stain.

Treatment

In the immunocompetent the disease is self-limiting. No effective therapy is available for the protracted, often fatal, diarrhoea in the immunocompromised although azithromycin or paromomycin may reduce the parasitic load and lessen the magnitude of the diarrhoea.

Prevention

See p. 180.

MICROSPORIDIOSIS

Epidemiology

Microsporidia are intracellular protozoans ubiquitous in the animal kingdom. They have emerged as an important cause of chronic, disabling diarrhoea in severely immunocompromised, human immunodeficiency virus (HIV)-infected patients throughout the world. Among the many genera and species of microsporidia, *Enterocytozoon bieneusi*, *Encephalitozoon cuniculi*, *Encephalitozoon hellem*, *Nosema corneum* and *Septata intestinalis* account for most human infections. Little is known about the mode of transmission.

Clinical features and complications

- Infection in non-HIV patients is quite rare and mostly involves the eye or central nervous system (CNS). Enteritis is the most common manifestation in HIV-positive patients, who are invariably severely immunocompromised (CD4 $< 100/mm^3$). Patients have chronic non-bloody watery diarrhoea, often associated with crampy abdominal pains, nausea and vomiting. Malabsorption and wasting invariably follow. *Enterocytozoon bieneusi* and *S. intestinalis* are the responsible organisms
- Cholangitis can complicate intestinal microsporidiosis
- Ocular microsporidiosis (keratoconjunctivitis) has been reported (due to *E. hellem*, *E. cuniculi* and *N. corneum*).

Diagnosis

Usually by EM demonstration of the parasite in small-intestinal mucosal biopsy specimen. Detection of the spores in faeces by a trichrome staining method is also proving successful.

Treatment

There is no effective therapy. Albendazole 400 mg twice daily often reduces diarrhoea for a period but it worsens again despite maintenance therapy.

ISOSPORA BELLI INFECTION

Epidemiology and clinical features

This is another coccidian protozoan parasite which causes chronic diarrhoea in immunocompromised individuals. Infection is found in the tropical environment and *Isospora* is an important cause of chronic diarrhoea among HIV-infected patients in subsaharan Africa. Immunocompetent individuals may suffer from self-limiting diarrhoeal illness. Mode of transmission is not clear, but may be both person-to-person and animal-to-person, or via contaminated water.

Diagnosis and treatment

Stool microscopy is the usual method. Co-trimoxazole controls diarrhoea by eradicating the parasite.

CYCLOSPORA INFECTION

Cyclospora is a newly recognised parasitic cause of diarrhoea in humans. It was previously known as CLB (cyanobacterium-like or coccidian-like body) and is probably a coccidian parasite.

Epidemiology

Cyclospora-associated diarrhoea has been reported in residents of and travellers to many areas of the world, including Europe, the Americas, India, Nepal and South-East Asia. Travellers to endemic areas are more prone to develop illness than native residents.

Clinical features

Diarrhoea is often prolonged for several weeks with flatulence, bloating, dyspepsia and weight loss. These symptoms are commonly protracted in AIDS patients.

Diagnosis

This is established by demonstrating *Cyclospora* oocysts in faeces by microscopy. These are acid-fast round bodies, larger in diameter than cryptosporidium. Jejunal aspiration and EM of jejunal biopsy may also reveal them.

Treatment

Co-trimoxazole seems to be effective and should be used.

AMOEBIC DYSENTERY

Epidemiology

This common, and frequently serious, disease caused by a protozoan, *Entamoeba histolytica*, occurs widely throughout tropical and subtropical Asia, Africa, the Middle East, and Central and South America. In temperate climate it has been prevalent among male homosexuals

- The parasite, which is a human pathogen, exists in two forms: cyst and amoeba (trophozoite). Humans become infected by ingesting the four-nucleated cysts in faecally contaminated food or water. Direct person-to-person transmission occurs among male homosexuals
- In the small intestine the cyst is digested, releasing four amoebae, which then migrate downwards and live on the colonic surface and multiply by binary fission. Encystation occurs in the left side of the colon where the faecal contents are more solid
- Cyst forms are more resistant to drying and can survive outside the human body for a long time, particularly in moist conditions. The amoe-

bic forms do not cause infection, if ingested, as they are rapidly destroyed by gastric acid and enzymes. The IP is variable.

Pathogenesis

Usually the motile amoebae live harmlessly on colonic mucosa but sometimes they become invasive, causing mucosal ulceration. Invasive amoebae characteristically ingest red cells. It has been suggested that the invasive strains possess distinctive isoenzyme motility patterns (zymodemes) but a consensus is lacking.

Clinical features

• The disease presents gradually over a period of 1–3 weeks, with worsening diarrhoea containing blood and mucus and accompanied by cramp-like abdominal pains and variable pyrexia

• The symptoms usually persist for several weeks before subsiding, even if no treatment is given. However, in such cases, relapses are common, occurring irregularly for months or years.

Complications

Fulminant colitis: in patients who are immunocompromised, pregnant or otherwise debilitated and malnourished

Amoeboma: sometimes a segment (usually ileocaecal) of the colon may become diffusely thickened from chronic inflammation, presenting clinically as a palpable and tender abdominal mass

Chronic amoebiasis: in endemic areas, findings of *E. histolytica* cyst are not uncommon in individuals with intermittent, non-bloody but mucoid diarrhoea. Some of these patients are genuinely suffering from low-grade invasive amoebic infection, but most are not

Amoebic liver abscess: see p. 208.

Diagnosis

• Sigmoidoscopy may suggest the diagnosis by revealing shallow ulcers scattered over normal-looking rectal mucosa. However, in more severe infections, there may be diffuse haemorrhagic colitis

• The diagnosis is confirmed by the finding of motile amoebae, containing ingested red cells, in freshly voided faecal specimens. Amoebae may be seen in the rectal biopsy specimens

• In invasive intestinal amoebiasis serum anti-amoeba antibodies are usually elevated. Serology does not distinguish between active disease and asymptomatic carriage in patients with previous active disease.

Treatment

• Metronidazole is the drug of choice for active disease (7 days). This is followed by a 10-day course of diloxanide furoate to further destroy any amoebae in the gut

• Asymptomatic cyst excretion should be treated with diloxanide furoate in non-endemic countries, to prevent future disease.

Prevention

See p. 180.

Food poisoning

A wide range of illnesses may occur through ingestion of food which has been contaminated with injurious substances, but the term 'food poisoning' is usually restricted to sharp illnesses occurring soon after eating a contaminated meal. A cause-and-effect relationship is usually established only when more than one person is involved, all sharing a common meal. This then becomes an outbreak.

Table 11.2 outlines the salient features of those important food-poisoning types which are related to microbial contamination of food. Chemical causes are outside the remit of this book.

Investigation of outbreaks of food poisoning

• Speedily review all suspected cases to try to determine time, place and everybody at risk

• Obtain as complete a list as possible of all foods eaten

• Obtain samples of as many foods as may still be available, and keep under refrigeration

• Determine on clinical grounds the most likely aetiological agents and alert the investigating laboratory accordingly. The IP is often a useful guide: short periods of less than 6 hours suggest ingestion of preformed toxin (e.g. *S. aureus*, *B. cereus*, shellfish poisoning), whereas IP greater than 12 hours, presence of fever and continuing symptoms beyond 24 hours suggest actual infection by living organisms

• Institute infection control proceedings if institutions like hospitals, nursing homes, etc., are involved

• Obtain samples of vomit and faeces and have them examined for bacterial and viral causes

• Follow carefully the food chain from supply to kitchen to table, to determine any unsafe practices

• Consider a case-control or cohort study to analyse the association between the suspected foods and disease.

MICROBIAL FOOD POISONING

Type	Usual source	Incubation period (hours, unless otherwise stated)	Clinical features
Salmonella	Poultry, eggs, meat, milk, food contaminated by human carriers	12–48 (range 6–96)	Diarrhoea, abdominal pain, vomiting and fever
Campylobacter	Poultry, milk	2–5 days (range 1–11 days)	Diarrhoea, abdominal pain, vomiting and fever
Clostridium perfringens	Meat dishes	12–36 (range 8–22)	Diarrhoea and abdominal pain, vomiting rare. Fever absent
Staphylococcus aureus	Food contaminated by human carriers left to stand, allowing bacterial multiplication and production of heat-resistant toxin	2–4 (range 1–7)	Nausea, vomiting, fainting, diarrhoea less common. Fever absent. Symptoms settle within 1 day
Bacillus cereus	Rice and other cereals	1–5	Predominantly vomiting, some diarrhoea and abdominal pain
		or 8–16	or predominantly diarrhoea
Vibrio parahaemolyticus	Imported seafood of Pacific origin	12–18 (range 12–48)	Abdominal pain, diarrhoea, sometimes nausea and vomiting, fever
Clostridium botulinum	Preserved or canned food	12–36 (range 2 hours–8 days)	Diplopia, ptosis, hoarse voice, cranial nerve paralysis, descending paralysis
Calcivirus, SRSV	Cockles, oysters, food contaminated by infected food handlers, irrigated crops	12–48	Vomiting and diarrhoea

Continued

Table 11.2. Microbial food poisoning: sources and clinical features

MICROBIAL FOOD POISONING

Type	Usual source	Incubation period (hours, unless otherwise stated)	Clinical features
Paralytic shellfish poisoning	Mussels, clams, oysters and scallops contaminated by marine algal dinoflagellates which produce a toxin	30 min	Dizziness, tingling, drowsiness, respiratory paralysis may develop rapidly
Diarrhoeic shellfish poisoning	Bivalve molluscs contaminated by marine algal dinoflagellates which produce a toxin	30 min–12 hours	Diarrhoea
Scrombotoxin fish poisoning	Scromboid (mackerel, tuna) and other non-scromboid fish (sardines, pilchards) allowed to spoil, leading to histamine production through putrefying bacterial action	2	Flushing, headache, nausea, palpitation
Ciguatera fish poisoning	Large fish of Pacific, Caribbean and Indian Ocean origin. These fish feed on smaller fish which have fed on algal dinofl-agellates with toxin production	Several hours	Nausea, vomiting, diarrhoea and neuromuscular disturbances

Table 11.2. *Continued*

BACILLUS CEREUS FOOD POISONING

Epidemiology and pathogenesis

Bacillus cereus is a spore-bearing aerobic organism, often present in small number in foods, and ubiquitous in soil. It is an important cause of food poisoning.

Inadequately reheated pre-cooked rice, for instance rice which has previously been left at room temperature and then used in preparing fried rice in Chinese take-away food shops, has been frequently implicated in the UK. Such storage encourages multiplication of heat-stable spores of *B. cereus* not destroyed by cooking. These elaborate two types of enterotoxin: one heat-stable (causes vomiting) and other heat-labile (causes diarrhoea). A variety of other foods have also been incriminated.

Clinical features and diagnosis

Two types of illness have been described. One has a 1–5 hour incubation period, and vomiting is the dominant feature. The other has a longer incubation period (8–16 hours) and abdominal pains are more common. Symptoms do not persist beyond 24 hours.

Diagnosis is by isolating the bacillus from faeces and food residue.

CLOSTRIDIUM PERFRINGENS FOOD POISONING

Epidemiology and pathogenesis

Type A strains of *C. perfringens*, both heat-resistant (commonest) and heat-sensitive, are a common cause of food poisoning in countries where cooking practices favour the multiplication of these organisms in food. Soil and the gastrointestinal tracts of animals and humans commonly contain these organisms. Heat-resistant spores contaminating food commonly survive cooking and later, during the slow cooking period, germination takes place with later production of enterotoxin (heat-labile) in the intestine. Inadequately heated or reheated meat and chicken dishes are commonly involved.

Clinical features, diagnosis and treatment

After an IP of about 12 hours, colicky abdominal pain and diarrhoea begin abruptly. Vomiting occurs occasionally. The illness is mild, lasting for less than 24 hours.

A severe form of disease with necrotising enteritis has been reported from Papua New Guinea, involving type C strains (pig bel).

Diagnosis is by semiquantitative estimation of *C. perfringens* in stools and food remains.

Attention to hydration is the only treatment.

Prevention

Meat dishes should be served hot soon after cooking or should be rapidly refrigerated. Bulk reheating should be avoided, otherwise storing temperature should be well above 60°C.

STAPHYLOCOCCAL FOOD POISONING

Epidemiology and pathogenesis

Staphylococcus aureus can multiply in cooked meats, milk products and rice to produce heat-stable enterotoxins which can cause nausea, vomiting, dizziness and fainting when consumed. Reheating of food destroys the organisms but not the toxin. Food handlers with septic skin lesions or skin/nasal carriage are often involved.

Clinical features, diagnosis and treatment

The illness begins abruptly after a short IP of 2–4 hours. Symptoms settle within 24 hours. Diarrhoea is infrequent. Confirmation of the diagnosis is by demonstrating the presence of enterotoxin-producing *S. aureus* in vomitus, faeces or food remains.

Attention to hydration is the only treatment.

Prevention

Education in food hygiene and exclusion of *S. aureus*-infected persons from food handling.

Typhoid and paratyphoid fevers (enteric fever)

Enteric fever is caused by *S. typhi* and *S. paratyphi* A, B and C, which belong to the genus *Salmonella*. They are primarily human pathogens and cause a prolonged febrile illness with bacteraemia and, initially, minimal gastrointestinal symptoms. In contrast, the other *Salmonella* are primarily animal pathogens and generally produce a diarrhoeal illness in humans.

Salmonella typhi, *S. paratyphi* A and B and some of the other *Salmonella* serotypes can be subdivided by phage typing. This is useful in epidemiological investigations.

Epidemiology

- The infections are most prevalent in South and South-East Asia, the Middle East, Central and South America and Africa, reflecting the poor standards of sanitation and water supply in these areas. A low level of endemicity exists in south and eastern Europe (mostly of paratyphoid B), and in most other developed countries, enteric fever is largely an imported illness. Paratyphoid C is rare everywhere
- Transmission is through food or water contaminated by the faeces or urine of a patient or carrier. Direct case-to-case spread is uncommon
- The IP is 10–21 days.

Pathogenesis

After penetrating the small-intestinal mucosa, the organisms travel to reticuloendothelial cell sites where they multiply, finally entering the blood stream in large numbers, marking the onset of fever. The Peyer's patches of the ileum are infected during this bacteraemia and also later through infected bile. They become inflamed and later during the 2nd or 3rd week of illness may ulcerate, causing haemorrhage and perforation. The liver and gall bladder are also involved. After recovery, symptomless biliary infection may persist indefinitely in the biliary and urinary tracts, particularly in the presence of pre-existing disease, leading to chronic faecal or urinary carriage.

Clinical features

Untreated typhoid fever is often a severe prolonged illness lasting for 4 or more weeks. Mild and inapparent cases also occur.

First week: slowly mounting remittent fever, headache, malaise, constipation, unproductive cough, relative bradycardia

Second week: continuous fever, mental apathy, diarrhoea, abdominal distension, rose spots in 30 per cent of cases (crops of 2–4 mm macules over lower chest and upper abdomen), splenomegaly in 75 per cent of cases

Third week: continuous fever, typhoid state (delirium, disorientation, drowsiness), gross abdominal distension, 'pea-soup' diarrhoea, complications

Fourth week: gradual improvement in fever, toxaemia and abdominal distension but intestinal complications may still occur.

After recovery, relapse may occur in up to 10 per cent of cases.

Complications

- Intestinal haemorrhage and perforation
- Myocarditis
- Neuropsychiatric: psychosis, encephalomyelitis
- Cholecystitis, cholangitis, hepatitis, pneumonia, pancreatitis
- Abscesses in spleen, bone or ovary (usually after recovery)
- Chronic carrier state: convalescent faecal carriage is not uncommon in patients treated with chloramphenicol, amoxycillin and co-trimoxazole, and about 3 per cent become long-term carriers (much less common following quinolone therapy).

Investigations and diagnosis

Definitive diagnosis requires positive blood on bone-marrow culture.

- Blood culture: highest positivity (80 per cent) during 1st week, declining thereafter. Prior antibiotic use reduces the frequency of positivity

• Bone-marrow culture: often remains positive despite antibiotic administration

• Stool and urine cultures: often positive from 2nd week onwards, diagnostic only if clinical picture is compatible

• Widal test: unreliable and often difficult to interpret in immunised persons, so rarely performed in Western countries

• A number of newer, more sensitive serodiagnostic tests such as Vi indirect immunofluorescent antibody test and detection of IgM antibody to *S. typhi* polysaccharide antigen are under investigation.

Treatment

• In hospital, patients should be nursed in isolation under enteric precaution

• Antibiotic therapy: ciprofloxacin orally or IV for 10–14 days in adults or third-generation cephalosporin, e.g. ceftriaxone, in children

• Adjunctive IV dexamethasone reduces mortality in severely toxic, obtunded patients

• 75 per cent of chronic carriers can be cured by 28-day courses of ciprofloxacin or norfloxacin. Cholecystectomy should be performed in patients only when symptoms of gall bladder disease warrant this

• Surgery is necessary for perforation, but haemorrhage is managed conservatively.

Prevention

• In non-endemic countries, epidemiological investigation to identify the source of infection is important if not acquired abroad

• After clinical recovery, 12 consecutive negative faeces and urine cultures should be obtained before declaring the patient free of infection. Food handlers should stay off work until then

• Individuals travelling to or living in highly endemic areas should receive typhoid vaccines. Three types are available, all giving about 70 per cent protection for 3 years:

  • killed whole-cell vaccine: two injections are necessary for primary course. Local and systemic side effects are common. It is cheap

  • Vi capsular polysaccharide: single injection, minimal local and systemic reactions. More expensive than killed whole-cell vaccine

  • Ty 21a live attenuated oral vaccine. three capsules over a 5-day period. Virtually free of side effects but costly. Not suitable for children under 5 years of age

• In typhoid endemic countries the most important measures are: provision of safe drinking water, safe disposal of excreta and public education on hygiene.

Abdominal tuberculosis

This is not uncommon in countries where tuberculosis is endemic, and in Britain is seen predominantly in immigrants from the Indian subcontinent.

Clinical features

Several clinical forms are recognised.

Tuberculous peritonitis: characterised by malaise, abdominal pain and distension. Ascites may be present or there may be palpable masses of matted omentum and loops of bowel. The diagnosis is by peritoneal biopsy, either percutaneously or during laparoscopy or laparotomy. The ascitic fluid is protein-rich from which tubercle bacilli may be cultured

Gastrointestinal tuberculosis: usually affects the ileo-caecal area causing narrowing, distortion and shortening. Endoscopic biopsy is diagnostic, but may be difficult to perform, so the diagnosis often depends on barium contrast images. Differentiation is from Crohn's disease. Evidence of tuberculosis elsewhere is helpful

Hepatic tuberculosis: low grade hepatic dysfunction is common in miliary tuberculosis which may present only as pyrexia of unknown origin (PUO), and liver biopsy will often show caseating granulomata. However, the occasional patient may present with jaundice, fever and an enlarged liver. Liver function usually shows a mixed cholestatic/hepatocellular picture. Liver biopsy is diagnostic.

Treatment

As for pulmonary tuberculosis (see p. 99).

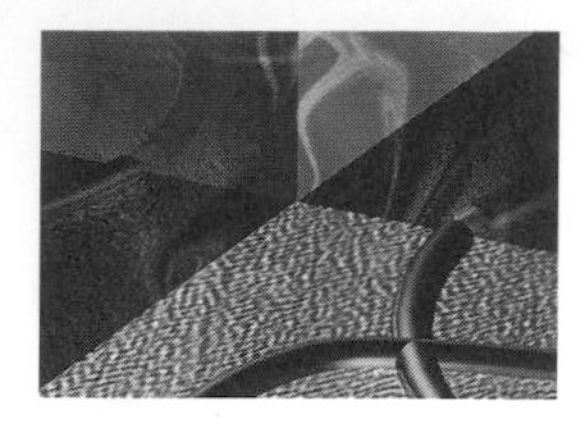

CHAPTER 12

Infections of the Liver and Biliary Tract

CLINICAL SYNDROMES

Liver abscess

Major presenting features: *fever, abdominal pain.*

Causes

Predisposing:

- biliary disease, gastrointestinal tract conditions (e.g. appendicitis, diverticulitis), systemic infection, contiguous infection
- residence in or travel to an endemic area for amoebiasis.

Microbiological:

- *STREPTOCOCCUS MILLERI*, ANAEROBES, *Escherichia coli*, other 'coliforms', *Enterococcus faecalis*, microaerophilic streptococci (often polymicrobial)
- *STAPHYLOCOCCUS AUREUS*, *Streptococcus pyogenes*, *Brucella suis*, actinomyces, *Salmonella typhi*, other *Salmonella*
- *ENTAMOEBA HISTOLYTICA*
- *Mycobacterium tuberculosis*, hydatid disease, *Candida*.

Distinguishing features

	Pyogenic liver abscess	*Amoebic liver abscess*
IP	1–16 weeks	1–12 weeks
Fever	high, hectic	low grade
Toxicity	may be marked	minimal or absent

	Pyogenic liver abscess	*Amoebic liver abscess*
Liver		
tenderness	usual	invariable, may be intercostal
swelling	uncommon	common
Jaundice	25%	5%
Clubbing	if chronic	never
Preceding events	intra-abdominal infection/surgery	dysentery 20%
Stool microscopy	normal	*E. histolytica* cysts/ trophozoites in 15%
Blood culture	positive 34%	negative
Abscess aspirate		
Gram	positive	negative
culture	positive	negative
trophozoites	negative	occasionally positive
Amoebic serology	negative	positive
Abscess number	multiple in 35%	rarely multiple

Complications

Extension: empyema, subphrenic abscess, hepato-bronchial fistula, skin, intrahepatic obstruction of major bile duct

Rupture: pericardium, peritoneum, bile duct, gastrointestinal tract

Metastatic: lung abscess, brain abscess

Others: pleural effusion, septicaemia (pyogenic abscess), secondary bacterial infection (amoebic liver abscess).

Investigations

- Full blood count (FBCt) (anaemia common, white cell count (WCCt) raised) and erythrocyte sedimentation rate (ESR) (raised)
- Biochemical profile (liver function tests (LFTs), albumin and urea/ creatinine). A raised alkaline phosphatase is often found
- Chest X-ray (CXR) (Fig. 12.1): raised hemidiaphragm anteriorly; pleural effusion/empyema; basal collapse; gas/fluid level in abscess; pericardial effusion (follows on from a left lobe abscess)
- Abdominal X-ray (calcification in the gall bladder or a hydatid cyst; gas/ fluid level in abscess)
- Imaging: ultrasound scan; isotope scan (technetium/gallium); computerised tomography (CT) scan
- Blood culture (positive in pyogenic abscess)

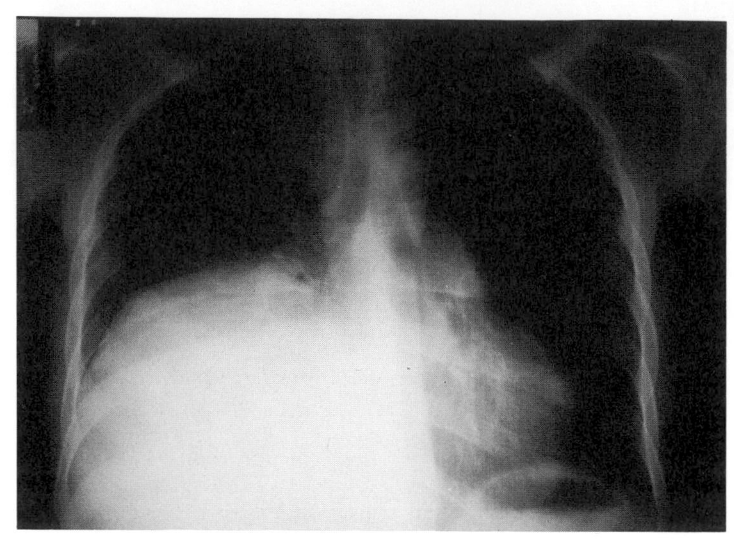

Fig. 12.1 Raised right hemidiaphragm from a large amoebic liver abscess.

- Abscess aspirate: microscopy (for trophozoites/cells); Gram stain; culture; cytology (malignancy can mimic an abscess)
- Serology for *E. histolytica*
- Stool microscopy (for amoebic trophozoites/cysts).

Differential diagnoses

- Hepatoma, metastases, cholangiocarcinoma, hepatic cyst.

Treatment

Pyogenic liver abscess

- Aspiration/drainage percutaneously under ultrasound scan (USS) guidance
- Antibiotics: metronidazole (for anaerobes), ampicillin (for streptococci, enterococci) and cefotaxime (for 'coliforms', *S. aureus*); or co-amoxiclav (antibiotics may need modification after culture results available)
- Formal surgery.

Amoebic liver abscess

- Metronidazole
- Aspiration/drainage percutaneously under USS guidance (if large abscess, left lobe, about to point, negative amoebic serology, doubt about amoebic aetiology)
- Luminal anti-amoebicide — diloxanide furoate (if stools are positive).

Prevention

- Prophylactic antibiotics for gastrointestinal surgery
- General travel advice concerning eating cooked food and boiling water.

Hepatitis

Major presenting features: *malaise, nausea/vomiting, jaundice.*

Causes

Acute hepatitis (transaminases 10–100 × normal)

- HEPATITIS A VIRUS (HAV), HEPATITIS B VIRUS (HBV), HEPATITIS C VIRUS (HCV), hepatitis delta virus (HDV), hepatitis E virus (HEV)
- Cytomegalovirus (CMV), Epstein–Barr virus (EBV)
- Yellow fever virus.

Acute hepatitis (transaminases 2–10 × normal)

- Leptospirosis, *Coxiella burnetii*, *Mycoplasma pneumoniae*, brucellosis, *Legionella pneumophila*, *Chlamydia psittaci*
- Pneumococcal pneumonia, Gram-negative bacteraemia
- Arboviruses
- *Treponema pallidum*, toxoplasmosis, *Candida*.

Chronic hepatitis (transaminases 1–10 × normal)

- HBV, HCV, HDV
- *Coxiella burnetii*, *M. tuberculosis*.

Distinguishing features

	Acute viral hepatitis	*Weil's disease (leptospirosis)*
IP	short (2–8 weeks): HAV, HDV, HEV long (1–6 months): HBV, HCV	10 days (6–15 days)
Onset	gradual	sudden
Risk history	contact, travel, shellfish (HAV/HEV); blood/ sexual contact, IVDA, institutions (others)	contact with animals/ contaminated water
Fever	normal/low	high
Headache	occasional	constant

	Acute viral hepatitis	*Weil's disease (leptospirosis)*
Chest symptoms	rare	not uncommon
Myalgia	mild	severe
Toxaemia	absent	marked
Conjunctivae	normal	suffused
Bleeding	rare	not uncommon
Liver failure	may occur (acute/chronic)	never
Proteinuria	absent	present
WCCt	normal	raised
Peak transaminase	100 × normal	2–5 × normal

Complications

Acute viral hepatitis

All types: fulminant hepatic failure, cholestatic hepatitis, relapsing hepatitis, aplastic anaemia, chronic fatigue syndrome

HBV, HCV, HDV: chronic active hepatitis, cirrhosis.

Chronic viral hepatitis

HBV, HCV, HDV: cirrhosis

HBV, HCV: hepatoma, vasculitis (polyarteritis nodosa (PAN), glomerulonephritis).

Leptospirosis

Renal failure, myocarditis, adult respiratory distress syndrome (ARDS) and disseminated intravascular coagulopathy (DIC) with fulminant progression.

Investigations

- FBCt, differential WCCt
- Clotting profile (prolonged clotting in liver failure)
- Biochemical profile (LFTs, albumin and urea/creatinine)
- Glucose (lowered in fulminant hepatitis)
- CXR (to exclude pneumonia, check for ARDS)

Acute viral hepatitis

- Serology
- Initial screen:
  - HBV surface antigen
  - HAV IgM antibody

- If initial screen negative:
    - HCV antibody
    - heterophile antibody test (Monospot)
    - serology for EBV, CMV, *C. burnetii*, *M. pneumoniae*, *Legionella pneumophila*, *Leptospira*, *Chlamydia*, *Toxoplasma* and syphilis
- If HBV surface antigen positive:
    - HBV IgM antibody (this confirms recent infection)
    - HBV e antigen/antibody (this determines infectivity)
    - HDV antibody.

Chronic viral hepatitis

- Serology:
    - HBV surface antigen
    - HCV antibody
    - HDV antibody (if HBV surface antigen positive)
- To determine activity of disease:
    - nucleic acid detection (HBV-DNA, HCV-RNA, HDV-RNA)
    - liver biopsy
- α-fetal protein (screen for hepatoma)
- Autoimmune screen (to exclude lupoid hepatitis).

Differential diagnoses

- Drugs, alcohol, autoimmune (lupoid) hepatitis, toxins, anoxic liver damage, Wilson's disease, haemochromatosis, α_1-antitrypsin deficiency, veno-occlusive disease, granulomatous hepatitis.

Treatment

Acute viral hepatitis

- Bed rest
- Vitamin K if prolonged prothrombin time
- Full intensive care for fulminant cases
- Liver transplantation.

Chronic viral hepatitis

- α-interferon up to 10 MU three times weekly for up to 6 months. Sustained response achieved in 35 per cent of HBV, 20 per cent of HCV, and 10 per cent of HDV infections
- Liver transplantation.

Prevention

Pre-exposure

- Passive immunisation: gammaglobulin for HAV (may be effective for HEV)
- Active immunisation:
  - HAV vaccine: booster at 6–12 months gives 10-year protection
  - HBV vaccine 0, 1, 6 months (serum antibody check at 6 weeks after last immunisation).

Post-exposure

- Passive immunisation:
  - gammaglobulin for HAV
  - hyperimmune HBV globulin for HBV.

General measures

See preventive measures listed under specific diseases, later in this chapter.

Cross-reference

Leptospirosis (see p. 311).

Cholecystitis and cholangitis

Major presenting features: *fever, right upper quadrant pain, jaundice.*

Causes

Cholecystitis

Predisposing:

- gallstones.

Cholangitis

Predisposing:

- gallstones, recent biliary surgery, biliary stricture, pancreatitis, *Clonorchis* infection, prior endoscopic retrograde cholangiopancreatography (ERCP), cholangiocarcinoma.

Microbiological:

- *ESCHERICHIA COLI*, OTHER 'COLIFORMS', ANAEROBES, *E. FAECALIS*, *S. typhi*, other *Salmonella*
- Kawasaki disease.

Distinguishing features

	Cholecystitis	*Cholangitis*
Preceding history	gallstones	gallstones, ERCP, surgery
Fever	normal/low	high, hectic
Rigors	absent	present
Toxicity	absent or mild	marked
Jaundice	rare	present
Abdominal tenderness	subcostal, on inspiration	upper abdominal
Palpable gall bladder	40%	uncommon
WCCt	mild/moderate rise	marked rise
Bilirubin	normal/mild rise	marked rise
Alkaline phosphatase	normal/mild rise	moderate rise
Positive blood culture	rare	50%
USS abdomen	patent, undilated CBD	dilated CBD

Complications

Cholecystitis: perforated gall bladder, biliary peritonitis, emphysematous cholecystitis, empyema of the gall bladder, cholangitis

Cholangitis: septicaemia, septic shock, perforation of the gall bladder, liver abscess, pancreatitis, empyema of the gall bladder.

Investigations

- FBCt, differential WCCt, ESR
- Biochemical profile (LFTs, albumin and urea/creatinine): mildly obstructive pattern usually seen
- Cardiac enzymes and electrocardiogram (ECG) (to exclude myocardial infarction (MI))
- Clotting profile (may be prolonged prothrombin time in cholangitis)
- CXR (to exclude pneumonia)
- Plain abdominal X-ray (for calcification, gas in gall bladder/CBD)
- USS of upper abdomen
- Blood culture.

Differential diagnoses

- Viral hepatitis, Weil's disease, liver abscess, pancreatitis, metastatic liver disease, pyelonephritis, peptic ulcer, MI, lower lobe pneumonia.

Treatment

- Intravenous (IV) cefuroxime (for 'coliforms'), metronidazole (anaerobes) with or without ampicillin (*E. faecalis*): or IV co-amoxiclav
- Surgery: immediate cholecystectomy (only if complications arise); elective cholecystectomy >3 months after acute cholecystitis
- ERCP and sphincterotomy if CBD gallstone obstruction.

Prevention

- Prophylactic antibiotics for biliary surgery and ERCP
- Therapy of gallstones.

SPECIFIC INFECTIONS

Hepatitis A

Epidemiology

HAV is a global disease of humans, particularly prevalent in developing nations:

- transmission is faecal–oral through contaminated food or water
- infectivity is greatest in the week before the prodrome and tails off after symptoms develop, becoming very low by the time jaundice develops
- anicteric infections are commoner in children (10:1) than adults (1:1)
- there are approximately 10,000 cases/year in the UK with an incidence of $15/10^5$ population; 5 per cent of cases are acquired abroad
- transmission rates are higher where there is poor sanitation, overcrowding, and amongst pre-school groups, male homosexuals and within institutions
- in developing countries most HAV acquisition occurs in childhood. In developed nations 20 per cent of young adults have serological evidence of past infection; incidence and severity rise with age
- large outbreaks have been described resulting from contaminated water, milk or food. Shellfish may become infected from sewage-contaminated seawater and become a vehicle for transmission
- lifelong immunity follows an attack
- the IP is 28 days (range 14–42 days).

Pathology and pathogenesis

HAV is an RNA enterovirus. Following ingestion, the virus reaches the liver via entry through the oropharynx or upper intestine. The virus is cytopathic but host-mediated immune responses also contribute to the acute hepatocellular damage. The findings on liver biopsy are nonspecific (focal necrosis, portal inflammation, ballooning, acidophilic bodies) and do not accurately distinguish HAV from other types of acute viral hepatitis. Severe hepatitis may be associated with massive necrosis.

Clinical features

Distinction of HAV from other types of viral hepatitis on clinical grounds is very difficult:

- the onset is usually gradual with low-grade fever, myalgia, upper abdominal discomfort, anorexia, nausea and vomiting. Smoking becomes distasteful
- after 3–6 days the urine becomes dark, the faeces pale and jaundice appears. This lasts for 1–2 weeks; other symptoms improve with its appearance
- arthralgia and rash occur in up to 5 per cent of cases
- hepatomegaly is found in nearly all patients and splenomegaly in 20 per cent.

Complications

Hepatic

- Fulminant hepatitis
- Cholestatic hepatitis
- Relapsing hepatitis.

Extra-hepatic

- Aplastic anaemia, haemolytic anaemia, thrombocytopenia
- Post-hepatitis syndrome (chronic fatigue syndrome).

Fulminant hepatitis occurs in 0.1 per cent of cases with a fatality rate of 20 per cent; this increases with age. Cholestatic jaundice is a common problem but always resolves and remains benign. Relapsing hepatitis 1–4 months after recovery associated with a return of hepatitis occurs rarely.

Diagnosis

A detailed clinical history and examination will suggest the diagnosis. Suspicion of hepatitis will come from:

- history of exposure (e.g. other cases in the family or school)
- very raised transaminases (e.g. alanine aminotransferase (ALT) or aspartate aminotransferase (AST) > 1000 IU/litre)
- prolonged prothrombin time.

Confirmation of HAV is through detecting HAV IgM antibody (positive for 12 weeks). The immunoglobulin G (IgG) antibody remains positive for life.

Treatment

- Bed rest is the mainstay of treatment
- Vitamin K if prolonged prothrombin time
- In fulminant cases, intensive care is necessary
- Liver transplantation can be life saving.

Prevention

- Good sanitation and water supply, together with personal hygiene, and avoidance of food or water likely to be contaminated
- Isolation of patients is unnecessary
- Children who are close contacts should be kept off school if they develop a febrile illness
- Shellfish should be cultivated in sewage-free waters
- Passive immunisation with normal immunoglobulin is effective and provides 3 months' protection. It is indicated for short-term travellers going to endemic areas, pregnant women and immunocompromised persons who are at risk of severe disease after close contact, and in health-care workers after occupational exposure
- Active immunisation with a killed vaccine produces excellent immunity. It is indicated for travellers going to endemic areas frequently or for >3 months, to abort outbreaks, and to protect health-care workers post-exposure, or pre-exposure where there is considered to be a significant occupational risk.

Prognosis

HAV is usually a mild illness. The fatality rate is 0.03 per cent of patients in those aged under 55 and 1.5 per cent in those age over 64 years.

Hepatitis B

Epidemiology

Worldwide, HBV is a major cause of disease through chronic liver disease and hepatoma.

- There are 10,000 new HBV infections/year acquired in the UK: the lifetime risk of hepatitis B is 5 per cent
- Anicteric infections are common (4:1)
- 5–10 per cent fail to resolve the infection and become carriers. This is more likely in persons with defective immunity
- The estimated carriage rate in the UK is 0.1 per cent. In certain areas of the world, the carriage rate may exceed 25 per cent (Pacific islands, Thailand, Senegal), and in others approximate 5–10 per cent (large areas

of the Indian subcontinent, South-East Asia, Africa and eastern Europe). The estimate is that nearly 200 million persons worldwide are carriers

- Transmission is by blood or body fluid through injection or exposure to mucous membranes. Infection can therefore be acquired from blood products, contaminated needles or other medical equipment, and life-style events such as tattooing. Infection may also be contracted sexually
- Newborn infants of infected mothers can acquire infection at birth. This is the major mode of transmission worldwide and is the major reason for high population carriage rates
- Infection and carriage rates are higher in closed groups where blood or other body fluids are injected, ingested or exposed to mucous membranes. Hence, institutionalised children with mental handicaps, haemodialysis patients, intravenous drug abusers (IVDAs) and male homosexuals have higher carriage rates (5–20 per cent). Outbreaks may occur in these groups and from infected surgeons and dentists
- Dual infection with the delta agent (HDV) may occur and is a particular problem amongst IVDAs
- The IP is 2–6 months.

Pathology and pathogenesis

HBV is a DNA virus possessing a surface coat (surface antigen) and an inner core (core antigen). In acute hepatitis, liver biopsy reveals varying degrees of hepatocellular damage and inflammatory infiltrate. HBV antigens are expressed on the surface of hepatocytes and there is T-cell-mediated cellular reactivity against these: this is presumed to be the major cause of hepatocyte damage. Patients with hypogammaglobulinaemia can develop acute hepatitis, indicating that antibodies do not play a major role in liver damage. In chronic hepatitis, varying degrees of histological activity may be seen. Chronic persistent hepatitis (CPH) is characterised by portal zone lymphocytic inflammation but with no evidence of bridging necrosis or disturbed architecture; these features are present in chronic active hepatitis (CAH). CPH is generally non-progressive whereas CAH may develop into cirrhosis or hepatoma. Chronic hepatitis is associated with chronic hepatitis B carriage and viral integration into the chromosome.

HBV surface antigen: this is found on the surface of the virus and on the accompanying unattached spherical particles and tubular forms. Its presence indicates acute infection or chronic carriage (defined as >6 months). Antibodies to surface antigen will occur after natural infection or can be elicited by immunisation

HBV core antigen: this is found within the core of the virus but is not detectable in the blood. IgG core antibody remains positive after infection indefinitely and is therefore a marker of past naturally acquired infection. IgM core antibody is useful in distinguishing HBV from another form of hepatitis in an HBV carrier (e.g. delta virus) and in the rare patient who clears their surface antigen quickly. IgM core antibody remains positive for 12 weeks

HBV e antigen: this is part of the core antigen. It is found in acute infection and in some chronic carriers. Its presence is a marker of underlying viral activity and infectivity. Antibody to this antigen indicates a lower level of infectivity in chronic carriers

HBV DNA: this parallels viral replication. It is found in acute hepatitis and chronic carriers with active disease.

Clinical features

The symptoms may be indistinguishable from HAV.

- The onset is usually insidious, with low-grade fever, anorexia, upper abdominal discomfort, nausea and vomiting, and distaste for cigarettes
- After 2–6 days, the urine darkens, the stools become paler and jaundice develops
- A syndrome of fever, arthralgia or arthritis, and urticarial or maculopapular rash occurs in 10 per cent of patients before the onset of jaundice. In children this may be pronounced and labelled papular acrodermatitis (Gianotti's syndrome)
- Smooth tender hepatomegaly is usual and splenomegaly occurs in 15 per cent of cases.

Complications

Hepatic

- Fulminant hepatitis
- CAH, CPH, cirrhosis
- Cholestatic and relapsing hepatitis
- Hepatoma.

Extra-hepatic

- Aplastic anaemia, haemolytic anaemia, thrombocytopenia
- Guillain–Barré syndrome (GBS), encephalomyelitis (rare)
- Post-hepatitis syndrome (chronic fatigue syndrome)
- Glomerulonephritis, vasculitis.

In 90 per cent of cases, the illness is benign and complete recovery ensues after 2–4 weeks. Fulminant hepatitis is more common with hepatitis B (1.0 per cent) than with hepatitis A (0.1 per cent) but remains rare; it is associated with infection with pre-S mutations in the genome of the HBV surface antigen and with acute co-infection and superinfection with delta virus (HDV). Chronic carriage occurs in 10 per cent and is associated with persistent (70 per cent) or active (30 per cent) chronic hepatitis on liver biopsy. Progression to cirrhosis or hepatoma occurs in 25–30 per cent of chronic carriers; this is more likely to occur in e antigen carriers. HBV has been incriminated in some patients with membranous glomerulonephritis and vasculitis.

Diagnosis (Fig. 12.2)

A detailed clinical history and examination will suggest hepatitis. Suspicion of *acute HBV* will come from the demonstration of:

- very raised transaminases (e.g. ALT or AST > 1000 IU/litre)
- prolonged prothrombin time
- consistent exposure history.

Confirmation is by demonstrating HBV surface antigen. Confirmation of *chronic HBV carriage* is by demonstrating:

- HBV surface antigen > 6 months after acute infection.

Confirmation of *chronic HBV hepatitis* is by demonstrating:

- raised transaminases > 6 months after acute infection
- liver biopsy evidence of CPH or CAH
- exclusion of other causes of chronic liver disease.

The level of infectivity and predicting underlying pathological activity can be gauged by determining HBV e antigen and antibody, and HBV DNA.

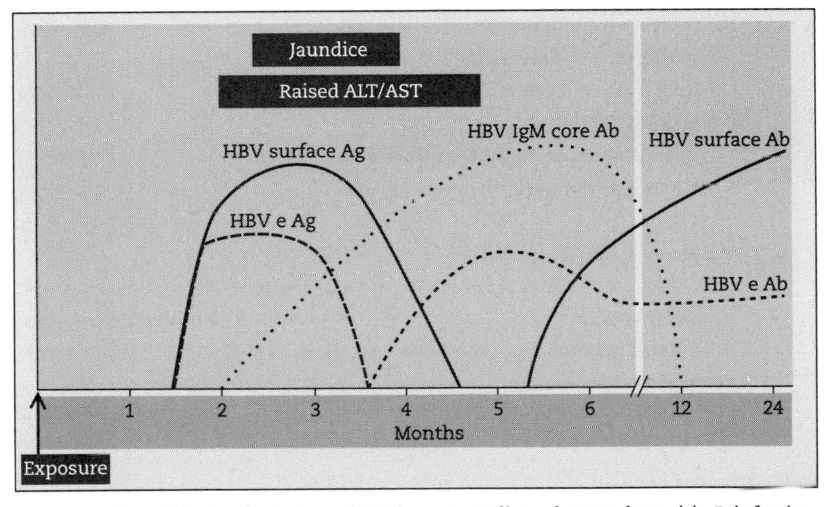

Fig. 12.2 Serological markers in uncomplicated acute hepatitis B infection. Ab, antibody; Ag, antigen.

Treatment

Acute hepatitis B

- Bed rest is the mainstay of treatment
- Vitamin K if prolonged prothrombin time
- In fulminant cases intensive care is necessary
- Liver transplantation is complicated by the probability of reinfection of the graft from extra-hepatic sites.

Chronic hepatitis B

Interferon thrice weekly for 6 months by subcutaneous injection has a success rate of 35 per cent with clearance of HBV e antigen if present initially. It is less likely to be effective in patients with cirrhosis, co-existent immunocompromising conditions (e.g. human immunodeficiency virus (HIV)), those infected vertically, and children.

Prevention

Two forms of protection are available: passive immunisation with hyper-immunoglobulin to hepatitis B and active immunisation with vaccine.

- Vaccine is indicated for newborn children whose mother is HBV surface antigen positive and for health-care workers post-exposure if previously unimmunised
- Hyperimmunoglobulin is indicated for newborn children of mothers carrying hepatitis B surface antigen who are also HBV e antigen positive or HBV e antibody negative. In the same instances it is also indicated for unimmunised health-care workers post-exposure
- Routine immunisation of at-risk groups is also important. These include all health-care workers, residents and staff of institutions for the mentally handicapped, and consorts and family members of a HBV e antigen positive carrier
- Elimination of high-risk persons as blood donors
- Screening of blood donors for HBV surface antigen.

Prognosis

Overall mortality of acute HBV is 1–3 per cent, but 25–30 per cent of patients with chronic carriage will have CAH, of whom 25 per cent will develop cirrhosis and/or hepatoma. The prognosis of these two conditions is very poor.

Hepatitis C

Epidemiology

HCV is a single-stranded RNA virus which cannot be cultivated. Through recombinant DNA technology, a diagnostic test has been devised which has identified hepatitis C as the major cause of what used to be termed non-A non-B (NANB) hepatitis.

- Studies in the 1970s and 1980s showed that the incidence of transfusion-associated NANB hepatitis was 10 per cent; subsequently, HCV was identified in 90 per cent
- The prevalence of HCV infection in the UK is 1:1400
- The major mode of transmission is through contaminated blood, most commonly via blood products or IVDA
- Sexual, vertical and occupational transmission do occur but much less frequently than with HBV; this reflects the lower concentration of HCV in the blood
- 70 per cent of anti-HCV positive persons have evidence of chronic hepatitis on biopsy but few have either signs or symptoms of liver disease
- 10–20 per cent of patients with chronic hepatitis are likely to develop cirrhosis within 5–30 years; 15 per cent of these will develop hepatoma
- Only 25 per cent are icteric; many infections are asymptomatic
- The IP is 8 weeks (range 4–26 weeks).

Pathology and pathogenesis

The findings on liver biopsy in acute HCV are non-specific (lobular disarray, hepatocyte degeneration, lymphocytic infiltration). More severe hepatitis may be associated with massive necrosis. As in the case of HBV, two broad classes of chronic hepatitis have been defined, CPH and CAH (see p. 219).

Clinical features

Clinically, HCV is indistinguishable from other causes of acute viral hepatitis.

- The onset is usually insidious, with low-grade fever, anorexia, upper abdominal discomfort, nausea and vomiting
- After 2–6 days, the urine darkens, the stools become paler and jaundice develops
- Prodromal arthralgia or arthritis may occasionally occur
- Smooth tender hepatomegaly is usual, and splenomegaly can occur.

Complications

Hepatic

- Fulminant hepatitis (1–2 per cent)
- CAH, CPH, cirrhosis
- Hepatoma.

Extra-hepatic

- Aplastic anaemia, agranulocytosis
- Cryoglobulinaemia.

HCV is one of the most common causes of chronic hepatitis. For the majority, disease progression is indolent and subclinical. HCV has been implicated in type II cryoglobulinaemia, which is a vasculitis.

Diagnosis

- Nonspecific features supporting HCV include:
  - mild hepatitis (transaminases 10 × normal as opposed to 100 × normal as is often seen with HBV)
  - prolonged prothrombin time
  - relapsing course
  - negative tests for HAV IgM antibody and HBV surface antigen
- Confirmation of hepatitis C is by detection of:
  - specific antibodies to HCV by enzyme-linked immunoadsorbent assay (ELISA). Newer tests have the advantage of fewer false-positive reactions and earlier positivity following infection
  - nucleic acid (HCV-RNA) by amplification (polymerase chain reaction (PCR))
- Confirmation of active hepatitis C is by detection of:
  - HCV-RNA. This distinguishes active from inactive infection
  - liver biopsy.

Treatment

For acute hepatitis, bed rest, and vitamin K if the prothrombin time is prolonged, are necessary. In fulminant cases, intensive care is needed. For chronic hepatitis, interferon-α thrice weekly for 3 months gives a 70 per cent transient response with improving LFTs and liver histology, but this is only prolonged in 20 per cent of cases. Only a few patients become HCV-RNA negative.

Prevention

- Elimination of high-risk persons as blood donors
- Screening of blood donors for HCV antibody
- Inactivation of HCV in blood products

- Use of synthetic plasma products produced through recombinant DNA technology (e.g. factor VIII)
- No vaccine is available.

Prognosis

Up to 20 per cent of patients with chronic hepatitis go on to develop cirrhosis within 5–30 years; 15 per cent of these will develop hepatoma. Life expectancy for these conditions is <5 years.

Hepatitis delta

Epidemiology

HDV is a defective RNA virus which can only replicate in cells already parasitised by HBV. It can therefore only present in a person co-infected with hepatitis B. Like HBV, HDV can cause acute and chronic liver disease. This may result from either simultaneous co-infection with HBV and HDV or from HDV superinfection in an HBV carrier. HDV:

- was originally described in Italy and is endemic in southern Europe, the Middle East, the Pacific islands and parts of Africa and South America. Transmission in this context is predominantly sexual
- has essentially been restricted to IVDAs and multiply transfused patients in western Europe and the USA; transmission is mainly through blood products. Infection is uncommon in other groups where HBV carriage is common
- rarely results in vertical transmission, HDV being more commonly associated with HBV e antibody positive status
- in the UK accounts for <1 per cent of all cases of acute hepatitis, increasing to 5 per cent where there is a history of IVDA. In chronic HBV carriers, it is found in 1–2 per cent of non-IVDA and one-third of IVDA patients. It is the cause of 6 per cent of all cases of chronic hepatitis, increasing to 25 per cent where there is a history of IVDA
- where contracted simultaneously with HBV and causing acute hepatitis, HDV is identical to acute HBV hepatitis although fulminant hepatitis appears more commonly; many cases are subclinical, as they are with HBV. Where acute hepatitis follows on from HDV superinfection on HBV carriage, both fulminant and chronic disease are more frequent
- in chronic carriers HDV is nearly always associated with chronic liver disease
- can be eradicated if HBV can be eliminated
- has an IP of 1–2 months.

Pathology and pathogenesis

Chronic carriage of HDV is associated with more severe liver damage on liver biopsy than HBV; rapidly progressive CAH or cirrhosis is observed. It is probable that like HBV (and unlike HAV and HCV), the virus is not directly cytopathic.

Clinical features

In acute co-infection with HDV and HBV, clinical features are no different from HBV alone (see p. 220).

Complications

- Fulminant hepatitis
- CAH, cirrhosis.

Hepatoma is not closely linked with HDV.

Diagnosis

Nonspecific features supporting acute HDV include HBV surface antigen positivity, and:

- very raised transaminases (e.g. ALT or AST > 1000 IU/litre)
- prolonged prothrombin time
- history of IVDA
- arrival from an endemic area
- knowledge of previous HBV carriage
- negative tests for HAV and HCV
- relapsing course and progression to chronic disease.

Confirmation of HDV is by detection of:

- HDV antibody (seroconversion may take place months after the clinical illness, so prolonged testing is necessary)
- HDV antigen
- HDV-RNA by amplification (PCR)
- HDV in liver biopsy.

Treatment

Acute hepatitis

- Bed rest, and vitamin K if the prothrombin time is prolonged
- In fulminant cases, intensive care is necessary.

Chronic hepatitis

Interferon-α thrice weekly for up to 16 weeks; half of the patients respond with a return of transaminases to normal and clearance of HDV-RNA from serum, although many patients relapse. Rarely, HDV may be cured if HBV is eliminated.

Prevention

- Exclusion of high-risk persons from blood donation
- Screening blood and blood products for HBV
- Inactivation of HDV in blood products
- Use of synthetic plasma products produced through recombinant DNA technology (e.g. factor VIII)
- Hepatitis B immunisation of those at risk
- No vaccine is available.

Prognosis

Chronic HDV infection is responsible for over 1000 deaths each year in the USA and patients are significantly more likely to require liver transplantation than are patients with HBV alone.

Hepatitis E

Epidemiology

HEV is a calicivirus and epidemiologically resembles hepatitis A. Sporadic cases have been reported from all countries but epidemic disease is mainly seen in developing nations. The first major outbreak due to HEV was in New Delhi in 1957, when there were 29,000 cases.

- Outbreaks have predominantly affected South-East Asia, Burma, Nepal, the former USSR, Mexico, Venezuela and north Africa. They are associated with gross contamination of water supplies, usually by sewage
- Secondary cases amongst household contacts do occur but less frequently than with HAV
- Transmission is by water and the faecal–oral route
- Both children and adults can be infected
- Approximately one-third of the cases of NANB hepatitis and half of those with non-A, non-B, non-C hepatitis are caused by HEV
- The seroprevalence in Europe and the USA is 1–2.5 per cent, compared to 10–15 per cent in South-East Asia
- There is an unusually high mortality of HEV during pregnancy (20–40 per cent)
- The IP is 6 weeks (2–9 weeks).

Pathology and pathogenesis

Through molecular cloning, three HEV proteins have been expressed and form the basis of immunoassays. This will allow better understanding of the virus.

Clinical features

These are indistinguishable from HAV (see p. 217).

Complications

Fulminant hepatitis probably occurs with the same frequency (0.1 per cent) and prognosis (20 per cent mortality) as in hepatitis A. The incidence and mortality is particularly high in pregnant women.

Diagnosis

Nonspecific features supporting a diagnosis of hepatitis E include:

- recent arrival from or travel to an endemic area
- negative tests for HAV IgM antibody, HBV surface antigen and HCV antibody
- no risk factors for hepatitis B or C
- no evidence of chronicity, with LFTs returning to normal.

Diagnosis of HEV can be confirmed by:

- detection of IgM and IgG antibodies to HEV
- detection of virus particles in stool by immune electron microscopy
- detection of HEV-DNA by amplification.

None of the HEV-specific investigations are routinely available at present.

Treatment

The management is the same as for hepatitis A (see p. 218).

Prevention

- Passive protection using immunoglobulin may have a role, but there are few data
- Travellers should be warned of the need to adequately cook food and boil water
- No vaccine is available.

Prognosis

Outside of complications in pregnancy, when fatal fulminant hepatitis is a major problem, HEV is a relatively benign infection.

CHAPTER 13

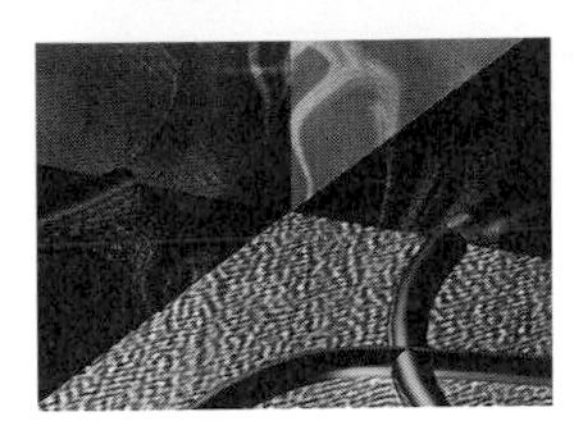

Infections of the Genitourinary Tract

CLINICAL SYNDROMES

Genital ulcers

Major presenting features: *ulceration, pain, lymphadenopathy*.

Causes

- HERPES SIMPLEX
- *TREPONEMA PALLIDUM*, *Haemophilus ducreyi*
- Lymphogranuloma venereum, granuloma inguinale (donovanosis).

Distinguishing features

	Primary syphilis	*Genital herpes*	*H. ducreyi (chancroid)*
IP	17–28 days	2–14 days	<10 days
Ulcer			
number	single	multiple	single/multiple
edge	erythematous, well-defined, rounded	erythematous, small, round	erythematous, ragged, undermined
induration	present	absent	absent
pain	absent	present	present
tenderness	absent	marked	marked
base	clean	clean	dirty
slough	clear	grey-white	necrotic

	Primary syphilis	*Genital herpes*	*H. ducreyi (chancroid)*
size where multiple	uniform	uniform	variable
Lymph-node			
swelling	moderate	nil to moderate	marked
tenderness	nil	marked	marked
bubo	absent	absent	frequent
Systemic symptoms	absent	if primary infection	absent
Diagnosis	dark ground illumination, serology	EM, viral culture, cytology, history of recurrences	culture

Complications

Primary syphilis: progression to secondary, latent, tertiary and quaternary disease

Herpes simplex: recurrences, aseptic meningitis, sacral radiculomyelitis, urinary retention, secondary bacterial infection

Haemophilus ducreyi: phimosis, inguinal buboes, auto-inoculation.

Investigations

- Ulcer swab:
  - dark ground illumination
  - Gram stain (*H. ducreyi*: 'schools of fish')
  - culture (viral, bacterial and fungal)
  - cytology (for viral inclusion bodies)
- Serology for:
  - Venereal Diseases Research Laboratory (VDRL), *T. pallidum* haemagglutination assay (TPHA) and fluorescent treponemal antibody (FTA)
  - *Chlamydia trachomatis* lymphogranulovenereum (LGV) serotypes
  - herpes simplex
- Biopsy.

Differential diagnoses

- Trauma, Behçet's syndrome, malignancy, Stevens–Johnson syndrome, fixed drug eruption.

Treatment

- Primary syphilis: procaine penicillin, or tetracycline if penicillin-allergic
- Herpes simplex: acyclovir
- *Haemophilus ducreyi:* co-trimoxazole, co-amoxiclav or ceftriazone

Prevention

- Herpes simplex: acyclovir can prevent recurrences
- Primary syphilis: progression can be prevented by adequate initial treatment
- Counselling as to safe sex, use of condoms, etc.

Urinary tract infection, urethritis and urethral syndrome

Major presenting features: *dysuria, frequency, urethral discharge.*

Causes

Urinary tract infection

Predisposing:

- lower urinary tract infection (UTI):
  - obstruction (prostatic hypertrophy, urethral valves or stricture)
  - poor bladder emptying (neuropathic, diverticula)
  - catheterisation/instrumentation
  - vesico-enteric fistula
  - sexual intercourse
- upper UTI:
  - vesico-ureteric reflux
  - obstruction (e.g. calculus, stricture).

Microbiological:

- *ESCHERICHIA COLI*, *PROTEUS* SPP., *KLEBSIELLA* SPP., OTHER 'COLIFORMS', *ENTEROCOCCUS FAECALIS*, *STAPHYLOCOCCUS AUREUS*, *STAPHYLOCOCCUS SAPROPHYTICUS*, *Staphylococcus epidermidis*, *Pseudomonas aeruginosa*
- *Mycobacterium tuberculosis*, *Schistosoma haematobium*, brucellosis.

Urethritis

- *CHLAMYDIA TRACHOMATIS*, *NEISSERIA GONORRHOEAE*, *UREAPLASMA UREALYTICUM*, *Mycoplasma genitalium*, *Trichomonas vaginalis*
- Herpes simplex.

Urethral syndrome

- *Chlamydia trachomatis*, *U. urealyticum*, *N. gonorrhoeae*
- Herpes simplex

- 10^{3-4}/ml pure growth of bacterium on midstream urine (MSU)
- *Lactobacillus* spp. overgrowth.

Distinguishing features

	Gonococcal urethritis	*Non-gonococcal urethritis*
IP	<1 week	>1 week
Onset	abrupt	insidious
Symptoms	constant, severe	intermittent, may remit
Dysuria and discharge	usual	one-third
Discharge	purulent	mucoid/mucopurulent
Gram stain	positive	negative

Complications

Lower UTI: pyelonephritis, septicaemia, epididymitis, prostatitis, chronic cystitis

Upper UTI: perinephric abscess, chronic pyelonephritis and scarring, septicaemia, renal failure, renal stones, ureteric stricture

Urethritis: epididymitis, urethral stricture, Reiter's syndrome, prostatitis, pelvic inflammatory disease, infertility, disseminated infection (*N. gonorrhoeae*), periurethral abscess.

Investigations

All

- MSU for microscopy, culture and sensitivity.

Upper UTI

- Blood culture
- Full blood count (FBCt), differential white cell count (WCCt) and erythrocyte sedimentation rate (ESR)
- Biochemical profile (liver function tests (LFTs), albumin and urea/creatinine)
- Chest X-ray (CXR) (to exclude pneumonia).

Urethritis and urethral syndrome

- Urethral swab: Gram and culture; *Chlamydia* antigen or culture
- High vaginal swab: microscopy for *Trichomonas*.

Differential diagnoses

Upper UTI

• Lower lobe pneumonia, cholecystitis, musculoskeletal pain, perinephric abscess, intraperitoneal abscess, pelvic inflammatory disease (PID).

Urethritis and urethral syndrome

• Vaginitis, interstitial cystitis, detrusor instability, trauma, urethral stricture, chronic prostatitis, PID.

Treatment

Upper UTI

• Intravenous (IV) antibiotics (e.g. cefuroxime).

Lower UTI

• Oral antibiotics, e.g. one of trimethoprim, co-amoxiclav or ciprofloxacin.

Urethritis

• *Neisseria gonorrhoeae*: amoxycillin, ciprofloxacin or azithromycin
• *Chlamydia trachomatis*: doxycycline, ciprofloxacin or azithromycin.

Prevention

• Prophylactic antibiotics for recurrent UTI (e.g. low-dose trimethoprim)
• Contact tracing is essential in identifying asymptomatic *N. gonorrhoeae* or *C. trachomatis* carriers who may develop complications, and in preventing reinfection and further transmission.

Pelvic inflammatory disease

Major presenting features: *lower abdominal pain, vaginal discharge, fever.*

Causes

Predisposing:
• young age, multiple sexual partners, intrauterine device, previous PID.
Microbiological:
• NEISSERIA GONORRHOEAE, C. TRACHOMATIS, *Mycoplasma hominis*, *U. urealyticum*
• *Escherichia coli*, other 'coliforms', *E. faecalis*, anaerobes
• *Mycobacterium tuberculosis*, actinomycosis.

Distinguishing features

	Gonococcal PID	*Non-gonococcal PID*
Onset	abrupt	insidious
Fever	$\geqslant$38°C	<38°C
Toxicity	may be marked	mild
Endocervical discharge	purulent	mucoid or mucopurulent
Abdominal pain/guarding	usually marked	less marked
WCCt ($\times 10^9$/litre)	$\geqslant$20	<20

Complications

• Recurrent PID (usually as a result of not treating sexual partner), chronic abdominal pain, ectopic pregnancy, infertility, tubal abscess, perihepatitis, multiple miscarriages, adhesions.

Investigations

• FBCt, differential WCCt, ESR
• Biochemical profile (LFTs, albumin and urea/creatinine)
• Blood culture (for gonococcal bacteraemia)
• MSU
• Endocervical swab: Gram stain; antigen detection (*Chlamydia*); culture (*N. gonorrhoeae*, *C. trachomatis* and general (including anaerobes))
• Ultrasound scan (USS) of pelvis
• Laparoscopy (and appropriate specimens as above).

Differential diagnoses

• Pyelonephritis, tubal pregnancy, ruptured ovarian cyst, appendicitis, miscarriage, endometriosis.

Treatment

Neisseria gonorrhoeae: amoxycillin (where low levels of resistance), ciprofloxacin or azithromycin
Chlamydia trachomatis: doxycycline, ciprofloxacin or azithromycin
Enteric flora: co-amoxiclav or ampicillin and metronidazole.

Because of the frequent finding of polymicrobial infections, combination therapy is advised, e.g. co-amoxiclav and doxycycline.

Prevention

• Contact-tracing of female sexual contacts of men with *N. gonorrhoeae* or *C. trachomatis* is important in preventing the development of PID and subsequent infertility

• Contact-tracing of male sexual contacts of women with PID is important in preventing infection and recurrent PID; the risk of infertility increases proportionately with every episode.

SPECIFIC INFECTIONS

Gonorrhoea

Epidemiology

Gonorrhoea is a mucosal infection of the columnar epithelium transmitted by sexual intercourse and caused by *N. gonorrhoeae*.

• Attack rates are highest in those aged 15–24 years who live in cities, belong to a low socioeconomic group, are unmarried or homosexual, or who have a past history of sexually transmitted disease (STD)

• The disease is highly transmissible, with infection rates of 50 per cent in women and 20 per cent in men after a single unprotected vaginal exposure

• 75 per cent of women are asymptomatic, compared to only 5 per cent of men

• Sites less frequently affected are the oropharynx, eyes (neonates), vulva, vagina or peritoneum (prepubertal girls) and perihepatic tissues; disseminated infection is rare

• The incidence has steadily increased between 1951 and 1980, since when it has fallen; there are approximately 12,000 cases in England per annum

• Severe systemic infections and ophthalmia neonatorum have become rare in developed nations

• Protective immunity does not develop and reinfections are common after re-exposure

• The IP is 2–7 days; the disease is more common in men.

Pathology and pathogenesis

Following attachment, gonococci penetrate into the epithelial cells and pass to the subepithelial tissues, where they are exposed to the immune system (serum, complement, immunoglobulin A (IgA), etc.) and phagocytosed by neutrophils. Virulence is dependent upon gonococci being able to attach and penetrate into host cells as well as being resistant to serum, phagocytosis and intracellular killing by polymorphonucleocytes. Factors conferring these virulence factors include pili, the outer membrane proteins (OMPs), lipopolysaccharide and IgA proteases.

Clinical features

In men

- Urethritis is associated with a purulent urethral discharge, dysuria and a hazy urine on first micturition
- Anorectal gonorrhoea is associated with:
  - perianal pain, pruritus, a mucoid or mucopurulent discharge or anal bleeding; infection is asymptomatic in 60 per cent
  - a distal visible (20 per cent) and histological (40 per cent) proctitis.

In women

Infection is associated with:

- vaginal discharge, dysuria, frequency, and back or low abdominal pain; in many, examination is normal
- cervicitis, urethritis or proctitis in decreasing order of frequency. Cervicitis is characterised by an erythematous friable cervix and mucopurulent discharge.

Complications

In both sexes

- Oropharyngeal disease (mild)
- Dissemination
- Perihepatitis (Fitz–Hugh–Curtis syndrome)
- Endocarditis, meningitis (both rare).

Dissemination occurs in 1–2 per cent and manifests with a large-joint arthritis or tenosynovitis, a sparse maculo-papular/pustular rash, usually on the extremities, and systemic symptoms. It is more common in women.

In men

- Prostatitis, epididymitis
- Urethral stricture, periurethral abscess or penile lymphangitis.

In women

- PID, sterility
- Bartholinitis/abscess formation
- Neonatal conjunctivitis in offspring.

PID embraces salpingitis, endometritis and adnexitis. It is suggested by lower abdominal pain and tenderness, fever, pain on cervical movement, and adnexal tenderness or masses. It is estimated that 10 per cent of patients who have bilateral salpingitis remain sterile. Neonatal conjunctivitis occurs in the offspring of infected mothers and is characterised by an acute severe bilateral conjunctivitis occurring 2–7 days after delivery.

Diagnosis

Diagnosis is by identification of the organism, either by microscopy of Gram-stained smears, culture and identification, or antigen detection using immunofluorescence or enzyme-linked immunoadsorbent assay (ELISA) methodology.

- Gram-staining of urethral discharge is positive in 95 per cent of men and of endocervical discharge in 60 per cent of women
- Culture is essential in women, and in rectal and oropharyngeal disease. Confirmation of identity can then be made by sugar fermentation or *N. gonorrhoeae*-specific antigen detection kits.

Treatment

Penicillin resistance may be mediated by plasmid-encoded penicillinase or chromosomally encoded changes to penicillin-binding proteins. It accounts for 3–5 per cent of isolates in the UK. Because of the importance of eradication to prevent further transmission, single-dose therapy is preferable, with either:

- oral amoxycillin (high dose 3 g) in areas of low-level penicillin resistance, or in pregnancy
- oral ciprofloxacin or intramuscular (IM) ceftriaxone
- IM spectinomycin should be used for patients who fail to respond to the above.

Prevention

- Sexual contacts must be examined, screened and treated appropriately
- There is no vaccine.

Syphilis

Epidemiology

Syphilis is a chronic, systemic, infectious disease characterised by clinically well-defined stages. It belongs to a family of spirochaetes (*Borrelia*, *Leptospira*, *Spirillum* and *Treponema*).

- *Treponema pallidum* cannot be cultured. Treponemal cardiolipin cross-reacts with host-tissue antigen which is the basis of the use of the VDRL in diagnosis

• Incidence has been steadily declining (1000 cases/year in the UK in the mid-1990s), congenital syphilis becoming exceptionally rare in developed nations. However, congenital and acquired disease remain important problems in developing countries

• Disease may be congenital or acquired, each being subdivided into early (duration <2 years from infection) and late stages (duration >2 years). Early acquired syphilis is further subdivided into primary, secondary and early latent stages. Late acquired syphilis is split into late latent, tertiary (reserved for benign gummatous syphilis) and quartenary (including all other manifestations of late syphilis)

• Patients may progress through stages sequentially or some clinical stages may be clinically inapparent. 40 per cent of patients with untreated latent syphilis go on to develop late-stage disease

• Transmission is by sexual contact, occasionally by intimate non-sexual contact, and rarely *in utero*, vertically at birth, through inoculation injury, or via blood transfusion

• Non-venereally acquired treponemes are phenotypically indistinguishable and include *T. pertenue* (yaws), *T. carateum* (pinta), and *T. pallidum* var. endemic syphilis (bejel). They do not affect the central nervous system and are geographically localised, whereas syphilis is worldwide

• Syphilis complicating human immunodeficiency virus (HIV) infection may be accelerated and atypical

• The IP is 17–28 days (range 9–90 days); patients are regarded as non-infectious 4 years after acquiring the illness.

Pathology and pathogenesis

Obliterative endarteritis occurs at all stages of the disease. It is associated with perivascular infiltration with macrophages and plasma cells in the primary chancre, hyperkeratosis in cutaneous secondary syphilis and central necrosis and granulomata in gummas. Treponemes enter the body through mucous membranes or abraded skin where the primary chancre develops. This heals spontaneously over 2–4 weeks and its disappearance coincides with the secondary or disseminated stage, when treponemes can be identified from the skin lesions as well as blood, lymph nodes and central nervous system.

Clinical features

The primary chancre (usually astride the coronal sulcus in men and on the vulva or cervix in women) has the following characteristics:

• it is a solitary, painless, non-tender, rounded lesion with a well-defined erythematous margin and an indurated clean base

• it is associated with rubbery, painless, non-tender inguinal lymphadenopathy
• it is an ulcer that heals without scarring by 6–8 weeks.

Complications

Secondary syphilis

The signs of secondary syphilis are numerous and may be preceded by mild constitutional symptoms of fever, and influenza-like symptoms. Characteristic features are:
• a maculo-papular rash (90 per cent of cases) which is non-pruritic, symmetrical and involves the entire body including the face, scalp, palms and soles. In the perianal area this is manifested as condylomata lata, in the mouth as mucosal ulcers (snail-track or mucous patches) and in the hair as patchy alopecia
• generalised and painless moderate lymphadenopathy (50 per cent), especially the epitrochlear glands
• uveitis, hepatitis (obstructive LFT pattern) and glomerulonephritis in a few patients.

Tertiary syphilis

Tertiary syphilis (skin, mucous membranes and bones) may develop 3 or more years after the primary stage and the characteristic lesion of this stage is the gumma:
• a cutaneous gumma is usually a punched-out ulcer with a 'wash-leather' base, central depigmentation and peripheral hyperpigmentation which heals to form a 'tissue-paper' scar
• destructive mucosal gummas, periostitis (sabre tibia), gummas of the liver and uveitis may also occur.

Quaternary syphilis

Quaternary syphilis has its major effects on the heart and nervous system.

Cardiovascular syphilis
• It occurs in 10 per cent of untreated patients
• It is primarily an aortitis with secondary aortic incompetence, coronary ostial stenosis (leading to angina) and eventually aneurysmal dilatation of the aorta
• It usually affects the ascending aorta ('aneurysm of signs' (sternal/rib erosion, right heart failure, lung collapse)) or the arch ('aneurysm of symptoms' (hoarseness, brassy cough, stridor, dysphagia)).

Neurosyphilis

Neurosyphilis is divided into meningovascular, parenchymatous and tabes dorsalis, although there is a great deal of overlap.

• It occurs in 20 per cent of untreated patients

• Meningovascular syphilis may appear as early as 5 years after initial infection and presents with an aseptic meningitis, cranial nerve palsies and occasionally hemiplegia

• Generalised paresis of the insane occurs later, at 10–20 years, with global cortical dysfunction leading to cognitive impairment, loss of inhibition, tremors, fits, Argyll Robertson pupils and, eventually, dementia

• Tabes dorsalis occurs at 15–35 years after primary syphilis. It is characterised by lightning pains, loss of sensory modalities (postural, temperature, deep and superficial pain), hypotonia, ataxia, Charcot's joints, neuropathic ulcers, loss of tendon reflexes and bladder disturbances.

Congenital syphilis

Congenital syphilis may result from infection *in utero* and manifest early, with snuffles, maculo-papular rash, osteochondritis (saddle nose, sabre shin), hepatosplenomegaly and anaemia; or late, with interstitial keratitis, frontal bossing, deafness, abnormal dentition and recurrent arthropathy.

Diagnosis

Primary and secondary syphilis

• Dark-ground illumination of a wet-mounted preparation taken from the chancre, condylomata or mucous lesion, showing characteristic spirochaetes

• Nonspecific (VDRL – detecting cardiolipin antibody) and specific (TPHA, FTA – detecting treponemal antibody) serological assays. In primary syphilis, the FTA precedes the TPHA, which in turn precedes the VDRL in becoming positive. In secondary syphilis, all three tests are strongly positive. The VDRL titre is an indicator of activity; specific tests remain positive for life. False-positive nonspecific tests are not infrequent and can occur transiently in many infections, as well as in systemic lupus erythematosus.

Latent, tertiary and quaternary syphilis

Latent syphilis is diagnosed by serology; tertiary and quaternary syphilis by a combination of clinical features and serology:

• up to 25 per cent of patients with late syphilis have negative nonspecific tests, although all have specific antibodies

• histopathology is diagnostic but this is rarely obtainable, except in skin and mucous membrane lesions.

In a patient with positive serology:

- calcification of a dilated ascending aorta with aortic incompetence is very suggestive of syphilitic aortitis
- an abnormal cerebrospinal fluid (raised protein, >5 white cells/mm^3, raised globulin) is suggestive of neurosyphilis.

Distinction of syphilis from non-venereal treponemal infection is impossible by serology.

Treatment

Primary and secondary syphilis

- Either procaine penicillin daily for 14 days, a single dose of benzathine penicillin G, or tetracycline or erythromycin for 14 days (if penicillin-allergic).

Tertiary or quarternary syphilis

- Procaine penicillin and oral probenecid for 21 days
- Corticosteroids are sometimes indicated in late-stage syphilis to reduce the risk of a Herxheimer reaction.

Prevention

- Half of the sexual contacts are infected; contact tracing is vital
- There is no vaccine.

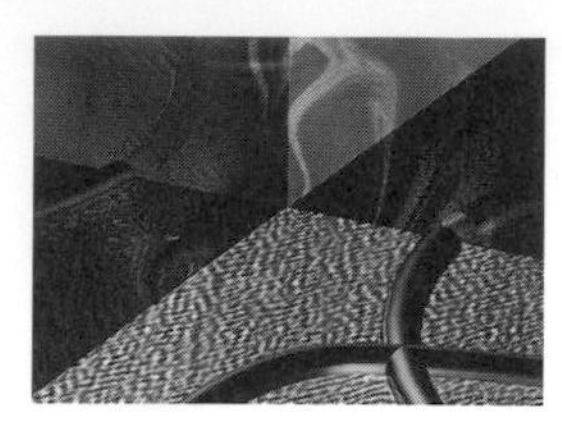

CHAPTER 14

Bone and Joint Infections

CLINICAL SYNDROMES

Osteomyelitis and infective arthritis

Major presenting features: *joint or bone pain, tenderness, redness with or without swelling, fever.*

Causes

Osteomyelitis

Predisposing:

• compound fracture, dental infections (jaw), post-cardiac surgery (sternal), soft tissue infection (pressure sores – sacrum)

• sickle cell disease, metal implant, diabetes mellitus (foot), peripheral vascular disease (foot).

Microbiological:

• *STAPHYLOCOCCUS AUREUS, STREPTOCOCCUS PYOGENES, HAEMOPHILUS INFLUENZAE* (TYPE B), *Salmonella typhi*, other *Salmonella*, *Escherichia coli*, other 'coliforms', *Pseudomonas aeruginosa*, anaerobes

• *MYCOBACTERIUM TUBERCULOSIS*, actinomycosis, hydatid disease.

Infective arthritis (native joint)

Predisposing:

• pre-existing arthritis (especially rheumatoid), intra-articular injection, metal implant, trauma, neighbouring osteomyelitis, corticosteroids, malignancy, intravenous (IV) drug abuse.

Microbiological:

• *STAPHYLOCOCCUS AUREUS, S. PYOGENES*, GROUP B β-HAEMOLYTIC STREPTOCOCCI, *H. influenzae* (type b), *Neisseria gonorrhoeae*, *E. coli*, other 'coliforms', *Salmonella*, *Neisseria meningitidis*, *P. aeruginosa*

• PARVOVIRUS, RUBELLA, hepatitis B, mumps, Epstein–Barr virus, adenovirus, influenza, arboviruses
• Lyme disease, rat-bite fever
• *MYCOBACTERIUM TUBERCULOSIS*, brucellosis
• REACTIVE ARTHRITIS (enteric pathogens, *Chlamydia trachomatis*), infective endocarditis, rheumatic fever (*S. pyogenes*).

Infective arthritis (prosthetic joint)

Predisposing:

• rheumatoid arthritis, corticosteroids, diabetes mellitus, obesity, advanced age, prior surgery
• wound haematoma or infection.

Microbiological:

• *STAPHYLOCOCCUS EPIDERMIDIS, S. AUREUS*, Gram-negative bacilli, anaerobes, other streptococci, diphtheroids, *Candida*.

Distinguishing features

	Septic arthritis	*Viral arthritis*
Onset	usually acute	gradual
Fever	moderate to marked	nil to moderate
Toxicity	moderate to marked	minimal
Number of joints	single	usually multiple
Pain, redness, swelling	marked	mild to moderate
Blood culture	positive in 50%	negative
Joint aspirate	turbid/purulent	clear
WCCt	$>$100,000/mm^3, neutrophilic	$<$100,000/mm^3, lymphocytic
Gram stain	often positive	negative
bacterial culture	positive	negative
Peripheral WCCt differential	raised, neutrophilic	normal/low, lymphocytic
Diagnosis	Gram, culture	serology

Complications

Osteomyelitis: chronic osteomyelitis (involucrum/sequestrum formation), septic arthritis, septicaemia, amyloidosis, sinus formation

Septic arthritis: reduced joint movement, destructive arthritis, osteoarthritis, osteomyelitis, contractures, septicaemia, sinus formation, avascular necrosis (hip)

Prosthesis infection: loosening, sinus formation, low-grade bacteraemia, chronic infection unless prosthesis removed.

Investigations

Osteomyelitis and arthritis

- Full blood count, differential white cell count, erythrocyte sedimentation rate
- Biochemical profile (alkaline phosphatase, calcium, phosphate, liver function tests, urea and creatinine)
- Blood culture
- Midstream urine (may be the source in elderly)
- X-rays of affected joint/bone (for lytic lesions, periosteal reaction: takes 2 weeks for abnormalities to develop).

Arthritis

- Autoimmune profile (rheumatoid factor, anti-nuclear factor, double-stranded DNA), urate; rheumatoid arthritis, systemic lupus erythematosus (SLE) and gout may mimic septic arthritis
- Ultrasound scan (especially for hip) to demonstrate any effusion
- Joint aspirate:
  - cell count, Gram stain, Ziehl–Nielsen stain
  - culture
  - crystals (to exclude gout, pseudogout)
- Serology
  - viral (mumps, rubella, parvovirus)
  - antistreptolysin O titre
- Arthrography (if chronic)
- Synovial biopsy (where negative cultures in chronic disease).

Osteomyelitis

- Isotope bone scan (technetium)
- Computerised tomography scan (benefit in identifying involucrum/sequestrum)
- Bone biopsy (as above).

Differential diagnoses

Osteomyelitis: osteosarcoma, benign bone tumour, sickle cell thrombotic crisis

Septic arthritis (native joint): rheumatoid arthritis, SLE, Still's disease, sero-negative arthritis (Reiter's syndrome, psoriatic, inflammatory bowel disease), gout, pseudogout, rheumatic fever, sarcoidosis, bursitis, haemarthrosis, irritable hip

Septic arthritis (prosthetic joint): mechanical loosening.

Treatment

Acute septic arthritis or osteomyelitis

- Antibiotics: initially IV (2 weeks) followed by oral (2–4 weeks)
  - *S. aureus*: flucloxacillin and initially either fucidin or rifampicin or gentamicin
- Aspiration (joint) or curettage (bone). Rarely, open surgical drainage of a joint is necessary
- Bed rest and immobilisation until joint/bone inflammation has subsided
- Then intensive physiotherapy.

Infected prosthetic joint

- Removal of prosthesis
- Antibiotics: initially IV (2–4 weeks) followed by IV or oral
  - *S. epidermidis*: vancomycin with or without another antibiotic to which the organism is sensitive.

Prevention

Antibiotic prophylaxis for implant surgery.

SECTION 3

Infection in the Compromised Host

CHAPTER 15

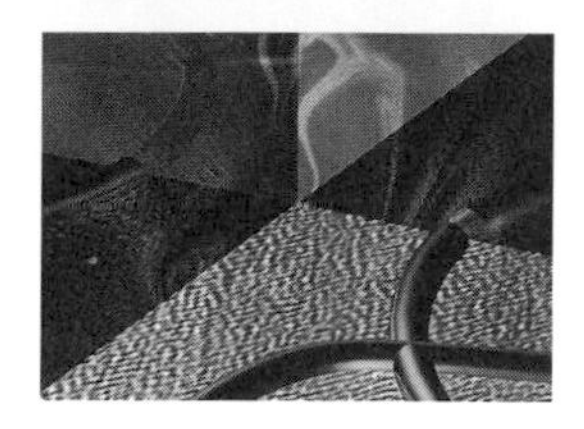

Infections Linked with Immunosuppression

ACQUIRED IMMUNODEFICIENCY SYNDROME

Acquired immunodeficiency syndrome (AIDS) is a highly fatal clinical syndrome which was first recognised in 1981. It is caused by infection with the human immunodeficiency virus (HIV) which produces progressive damage to the body's immune system, making it vulnerable to a wide range of infections and also to some forms of cancer. AIDS represents the late stage in the natural history of HIV infection.

HIV is a retrovirus belonging to the group of lentiviruses, and two distinct serological types are recognised – HIV-1 and HIV-2. Other closely similar lentiviruses cause diseases in animals, e.g. simian immunodeficiency virus.

Epidemiology

HIV infection has occurred in virtually all countries. Three distinctive waves of epidemic spread are recognised (see Table 15.1). The World Health Organisation (WHO) estimates there are currently about a million cases of AIDS worldwide and about 20 million people are HIV infected, of which more than 1 million are children, mostly in Africa. If unchecked, WHO predicts about 40 million infections by the year 2000, most of the rise occurring in Asia. Most infections are due to HIV-1. HIV-2 infection has occurred primarily in west Africa.

Transmission occurs only through the following routes:

• anal, vaginal or oral intercourse. The presence of genital ulcers facilitates transmission
• introduction of infected blood, blood products or tissues
• transmission to fetus from infected mother during pregnancy (13–40 per cent) and, uncommonly, during breast-feeding.

Saliva, tears, casual bodily contact and insect bites do not transmit infections.

THREE WAVES OF HIV EPIDEMIC

	First wave	Second wave	Third wave
Onset of epidemic	Late 1970s	1970s	Late 1980s
Countries involved	North America, Latin America, Western Europe, Australasia	Subsaharan Africa, parts of the Carribean Islands	South, South-East, East Asia, North Africa, Eastern Europe
Transmission (main groups)	Homosexual men, IV drug users, heterosexual people (more recently), blood products (now stopped)	Heterosexual	Travellers (initially), IV drug users, heterosexual people, blood products, unsterile needles
Incidence of AIDS cases	Plateauing	Rising	Still mainly asymptomatic
Incidence of new HIV infection	Falling	Plateauing/ falling	Rapidly rising

Table 15.1. Three waves of HIV epidemic

Pathogenesis and immunology

The manifestations of HIV disease are related either to the progressive destruction of immune function or to the damage caused by the virus itself to various organs, particularly the nervous system.

Immunosuppression is caused by preferential destruction (mechanism unknown) of a subset of thymus-derived (T) lymphocytes known as T-helper, T4 or CD4+ cells. Macrophages are also infected. As CD4+ cells decline in number, the body's cell-mediated immune defence becomes unable to protect it from invasive infections by viruses, bacteria, parasites and fungi. Congenitally infected babies additionally develop defective humoral immunity as CD4+ cells regulate B-cell maturity. Recurrent pyogenic bacterial infections are thus more common in HIV-infected children than in adults.

Immunosuppression also encourages certain forms of cancers, e.g. Kaposi's sarcoma, lymphoma, cervical and anal neoplasia. HIV also damages the nervous system directly, either during primary infection (acute meningitis or encephalitis) or during the late stage of the disease (subacute encephalitis, myelopathy, peripheral neuropathy).

Clinical features

The clinical course of HIV infection can be divided into several phases depending on the degree of immunosuppression.

Primary infection

This may be asymptomatic (more common), or symptomatic, when there may be fever with or without rash or a glandular fever-like illness (with mononucleosis) or, rarely, acute meningitis or encephalitis. These symptoms coincide with the development of HIV antibodies in the blood (seroconversion illness), which generally occurs within 3 weeks to 3 months of acquiring infection. CD4+ cells are often low initially but become normal on recovery.

Asymptomatic infection or persistent generalised lymphadenopathy (PGL)

After recovery from the acute infection, the patient remains asymptomatic for a prolonged period, usually lasting for 6–8 years. Some patients may have persistently enlarged, painless lymph glands in the neck, axillae and groins (PGL). The CD4+ lymphocytes remain normal ($>500/mm^3$) for a long time but always decline to well below $500/mm^3$ before the patient develops symptoms.

Early disease

Symptoms are usually mild and are related to skin and mucus membranes: seborrhoeic dermatitis, pruritic folliculitis, aphthous mouth ulcers, fungal skin and nail infections, oral hairy leucoplakia (white ridged patches on the lateral aspects of the tongue and on the buccal mucosa) and herpes zoster. CD4+ count is usually below $350/mm^3$.

Intermediate disease

The patient is now more immunocompromised, with CD4+ count nearer $200/mm^3$ (often much less). Common problems are persistent/recurrent fever or diarrhoea for more than 1 month, oropharyngeal candidiasis, persistent or recurrent vulvo-vaginal candidiasis and some weight loss.

Advanced disease (AIDS)

The patient is regarded as having AIDS when he or she develops a specified opportunistic infection or a secondary cancer. Definition of AIDS varies from country to country. Table 15.2 shows the UK AIDS-indicator diseases. The patient is severely immunocompromised with CD4+ lymphocytes almost invariably below $200/mm^3$. A patient may go

AIDS-INDICATOR DISEASES IN THE UK

- Bacterial infections recurrent (in a child <13 years)
- Candidiasis of oesophagus or respiratory tract
- Cervical cancer, invasive
- Coccidioidomycosis, extrapulmonary
- Cryptococcosis, extrapulmonary
- Cryptosporidial diarrhoea (>1 month)
- Cytomegalovirus disease not of spleen, liver or nodes
- Dementia (AIDS dementia)
- HSV ulcer (>1 month), bronchitis, pneumonia, oesophagitis
- Histoplasmosis, extrapulmonary
- Isosporiasis diarrhoea (>1 month)
- Kaposi's sarcoma
- Lymphoma, Burkitt's immunoblastic or primary cerebral
- Mycobacteriosis, disseminated or pulmonary tuberculosis
- PML
- *Pneumocystis carinii* pneumonia
- Pneumonia, recurrent within 1 year
- Recurrent *Salmonella* septicaemia
- Toxoplasmosis of brain
- Wasting syndrome due to HIV

Table 15.2. List of diseases which indicate AIDS in the UK

through the mild and/or intermediate disease states before developing AIDS, or remain asymptomatic before developing an AIDS-indicator disease. (Kaposi's sarcoma and lymphoma may occur at higher levels of CD4+ count.)

Laboratory tests

The following tests are useful for monitoring disease progression:

• CD4+ counts are currently the most reliable single test for this purpose. Steadily declining count and counts below 200/mm^3 signify poor prognosis

• Low CD4 : CD8 ratio, raised β_2-microglobulin and raised neopterin, and p24 antigenaemia also indicate unfavourable prognosis.

Specific infections

PNEUMOCYSTIS CARINII PNEUMONIA

Epidemiology

This is the most common AIDS-diagnosing condition in the Western world, although its incidence is now decreasing because of wide use of anti-*P. carinii* pneumonia (PCP) prophylaxis. It is caused by a parasite, *P. carinii* (recent RNA study suggests it may be a fungus). Asymptomatic infection is highly prevalent among the general population, and occurs from an early age. Clinically manifest illness occurs when reactivation occurs in highly immunocompromised patients. Previously, the disease was rare and seen only in patients receiving immunosuppressive therapy. The disease is rare when CD4+ counts are above 200/mm^3.

Clinical features

Fever, non-productive cough and progressive shortness of breath are common presenting symptoms. Chest X-ray (CXR) characteristically shows bilateral diffuse, interstitial infiltrations (Fig. 15.1). Hypoxaemia is common.

Diagnosis

PCP can usually be diagnosed if clinical and radiological findings are characteristic, otherwise confirmation should be obtained by demonstrating *P. carinii* in induced sputum (less reliable: <50 per cent success rate) or broncho-alveolar lavage fluid (over 90 per cent success rate) and/or in transbronchial biopsy (most reliable). Abnormal gallium isotope lung scans, raised lactate dehydrogenase (LDH) and demonstration of poor lung diffusion capacity are helpful but not diagnostic.

Treatment

Co-trimoxazole (120 mg/kg/day) IV or orally, or IV pentamidine (4 mg/kg/

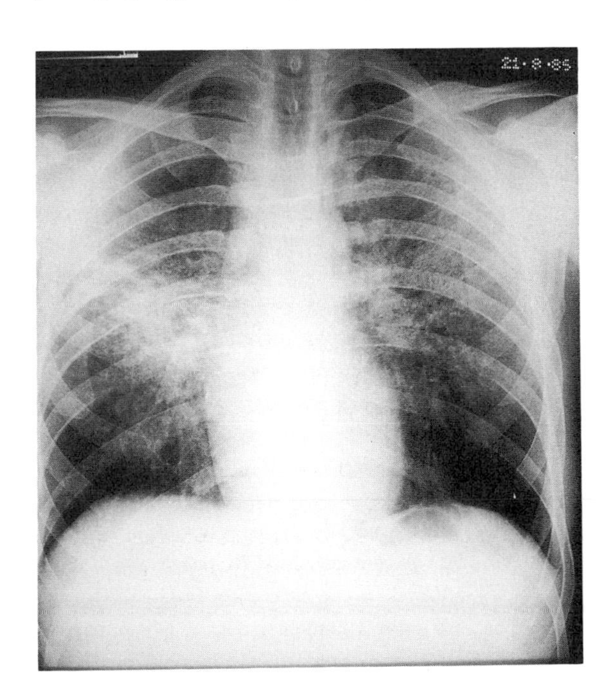

Fig. 15.1 *Pneumocystis carinii* pneumonia in an HIV-infected person.

day), for 14–21 days are the drugs of choice (the former is used most commonly). Adverse reactions are common with either drug (drug reaction and haematological side effects with co-trimoxazole, and hypotension, hypokalaemia, hypocalcaemia with pentamidine) necessitating a change to the other drug.

Treatment failure may occur with either agent, and second-line drugs are: trimetrexate with leucovorin, piritrexime or difluororthrithrin or atovaquone. In milder cases, primaquine/clindamycin or daily nebulised pentamidine may be used as primary therapy. Prednisolone reduces mortality and should be given to patients with P_{O_2} below 70 mm. Ventilatory support should be considered in suitable cases of severe PCP.

Prognosis

With prompt therapy, the mortality in first episodes of PCP is quite low, but in relapsed cases or in failure of first-line therapies, mortality is high. Complication such as pneumothorax occurs commonly in such cases.

Prevention

All patients with CD4+ lymphocytes below 200/mm^3 should receive anti-PCP prophylaxis with either co-trimoxazole (960 mg/day), dapsone (50 mg/day) or nebulised pentamidine (600 mg fortnightly).

CEREBRAL TOXOPLASMOSIS

In patients with prior latent infection (common in the general population) reactivation may occur, presenting usually as cerebral abscess. Computerised tomography (CT) scan characteristically shows multiple, hypodense shadows which ring-enhance with contrast (Fig. 15.2). Single lesions are difficult to differentiate from cerebral lymphoma by CT, and therapeutic trial of anti-*Toxoplasma* drugs with serial CT scans is indicated in such cases if brain biopsy is not contemplated.

The treatment of choice is pyrimethamine for 6 weeks. Folic acid supplements are necessary. Side effects are common, particularly sulphonamide allergy when clindamycin can be used or atovoquone alone. Maintenance therapy (pyrimethamine and sulphonamide, or co-trimoxazole, or clindamycin and trimethoprim) is essential to prevent relapse.

A negative *Toxoplasma* serology rules out cerebral toxoplasmosis with 90 per cent certainty. Demonstration of either intrathecal production of antibody or cerebrospinal fluid (CSF) polymerase chain reaction (PCR) are experimental procedures which may help. There is evidence that co-

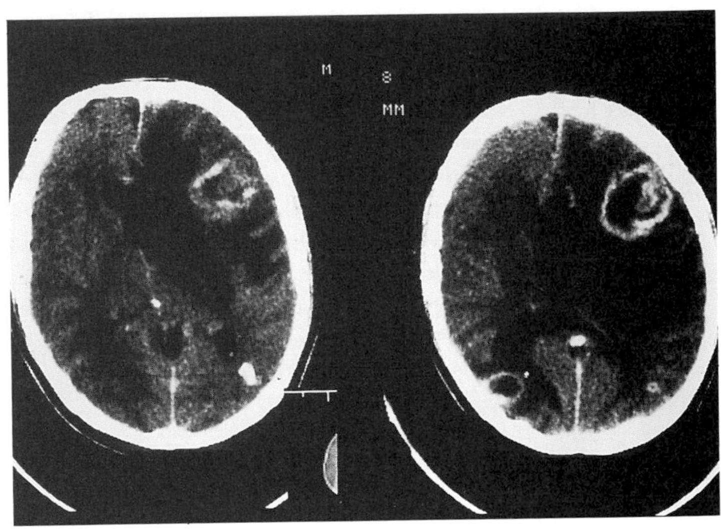

Fig. 15.2 Cerebral toxoplasmosis in an HIV-infected person before (right) and after (left) 2 weeks of treatment.

trimoxazole given for PCP prophylaxis in patients with serological evidence of past infection offers some protection against cerebral toxoplasmosis and should be used in all immunosuppressed HIV patients with serological evidence of previous *Toxoplasma* infection.

DISSEMINATED MYCOBACTERIUM AVIUM COMPLEX INFECTION

Mycobacterium avium complex (MAC) belongs to the group of atypical mycobacteria which are environmental organisms commonly found in water and soil. In the immunocompetent they are rare causes of pulmonary disease and cervical lymphadenitis. In severely immunocompromised HIV patients (CD4+ cell count below 50/mm^3) MAC is common. Fever and sometimes diarrhoea are usually the only symptoms, rarely pulmonary and neurological manifestations are present. Diagnosis is by isolation of MAC from blood, bone-marrow or liver biopsy specimen. Importance of isolation from stool or sputum is uncertain.

MAC is resistant to most standard antituberculous drugs. Various multi-drug combinations involving rifabutin, clofazimine, ethambutol, ciprofloxacin and clarithromycin often relieve symptoms and possibly prolong survival, but which drugs, how many and for how long, are subjects of debate. Rifabutin reduces the incidence of disseminated MAC in patients with CD4+ <50/mm³ when used as prophylaxis.

CROSS-REFERENCES

Candidiasis (see p. 261).
Histoplasmosis (see p. 265).
Coccidioidomycosis (see p. 266).
Cryptococcosis (see p. 266).
Cytomegalovirus (see p. 267).
Isosporiasis (see p. 197).
Microsporidiosis (see p. 196).
Progressive multifocal leucoencephalopathy (see p. 133).

HIV-related cancers

KAPOSI'S SARCOMA

Kaposi's sarcoma is a form of vascular neoplasm which used to be seen only in localised form in certain parts of Africa but occurs commonly in HIV-associated immunosuppression. Homosexual or bisexual men are mostly affected; skin and the alimentary mucosa are affected most commonly. Lesions are usually multiple, purple, raised nodules of several centimetres diameter. In the early stages, the disease is asymptomatic but later oedema and pain/discomfort are common, particularly involving limb, periorbital, palatal and gingival lesions. Visceral involvement other than the gut is uncommon and the lung is affected predominantly.

Treatment is necessary for cosmetic reasons, for alleviation of local symptoms and when there is major visceral involvement, i.e. lung. Localised lesion should be treated with radiotherapy but patients with widespread disease or visceral involvement require systemic therapy. Cytotoxic drugs (vincristine, vinblastine, adriamycin, etoposide) have been tried and the combination of vincristine and bleomycin is currently popular in Europe. Liposomal doxorubicin is equally effective but less toxic. Treatment is essentially palliative, and side effects often prevent long-term therapy.

B-CELL LYMPHOMA

This is a form of non-Hodgkin's lymphoma which originates from B-lymphocytes, similar to those already known to occur in immunosuppressed transplant recipients. A viral cause (especially Epstein–Barr virus (EBV)) has been suspected. In AIDS the disease is typically extranodal, involving the gastrointestinal tract, central nervous system (CNS), bone marrow and liver. Soft tissues and paranasal sinuses can be involved. Treatment is unsatisfactory.

CERVICAL AND ANAL CARCINOMA

Human papilloma virus (HPV) is causally linked to the development of cervical and anal cancers. There is evidence that HIV-induced immunodeficiency promotes the neoplastic changes in HPV-infected cervical and anal mucosa. Treatment is unsatisfactory; anti-cancer therapy does not seem to prolong life.

Neurological involvement due to direct HIV infection

Acute meningitis or encephalitis may develop during primary infection. Otherwise, neurological involvement occurs commonly in the late stages of AIDS and usually presents as one of the following.

Chronic HIV encephalopathy (AIDS dementia)

Forgetfulness and loss of concentration are early signs, followed by loss of balance and weakness of the legs. Dementia, with or without confusional states, develops later. CT scan shows cerebral atrophy (Fig. 15.3).

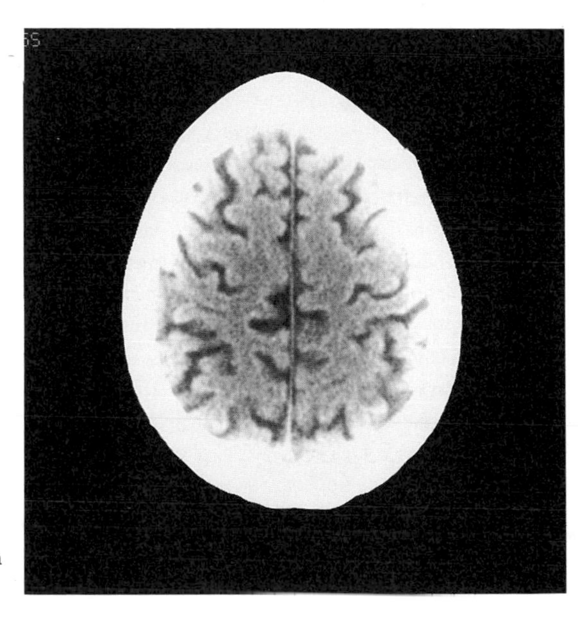

Fig 15.3 Cerebral atrophy in AIDS dementia.

Vacuolar myelopathy

This is usually seen in association with the preceding condition. Gradually progressive upper motor weakness of the legs, and paraesthesia, are the characteristic features.

Peripheral neuropathy

This is not AIDS-defining, and may occur without the presence of severe immunosuppression. It may take the form of weak muscles with diminished sensation and reflexes, or very distal painful neuropathy.

Autonomic neuropathy

This presents with bladder and bowel problems, and impotency.

Other manifestations of HIV disease

- Haematological: anaemia and thrombocytopenia (immune or due to defective production)
- Cardiomyopathy, glomerulonephritis, adrenocortical hypofunction, pancreatitis, polyarthritis, myopathy
- Infective: superficial mycosis, aspergillosis, *Penicillium marneffei* infection, leishmaniasis, listeriosis, genital herpes, molluscum contagiosum, bacillary angiomatosis
- Idiopathic diarrhoea and fever: may be chronic and disabling.

HIV disease in children

Congenitally infected children often develop symptoms within the first few years of life. Failure to thrive, oral thrush, frequent bacterial respiratory, ear, sinus or soft tissue infections, chronic parotid enlargement, hepatosplenomegaly and delayed milestones are often present. Opportunistic infections are seen as in adults, and lymphoid interstitial pneumonitis (bilateral lung infiltrate), hypoxaemia and clubbing of fingers/toes may develop.

Lymphoma and Kaposi's sarcoma are rare. Most children with symptomatic HIV infection during their first year die within the first 3 years. Others may remain asymptomatic for a number of years and the disease often behaves as in adults. Children infected through blood products also fall into this category.

Diagnosis

- HIV infection should be considered when symptoms are compatible with HIV disease and no satisfactory alternative explanation has been found. Infection is more often detected during voluntary testing of individuals presenting at testing clinics. Adequate pre- and post-test counselling are prerequisites of HIV testing

- Confirmation of HIV infection is usually achieved by demonstrating HIV antibody in blood. This usually develops within 3 months after infection. First-line tests are enzyme-linked immunoadsorbent assay (ELISA)-based but positive tests need confirmation by a second test, usually by the Western blot method (measures antibody production against viral components)
- Diagnosis of seroconversion illness: an antibody test is often negative in the early stages but p24 antigen is present in the blood, with antibody appearing later (Fig. 15.4).

Diagnosis of congenital HIV infection

Maternally derived antibody is often present for up to 18 months postpartum, so during this period a positive HIV antibody test is valueless. Laboratory confirmation of HIV infection can be achieved by one of the following means:

- persistence of HIV antibody beyond 18 months

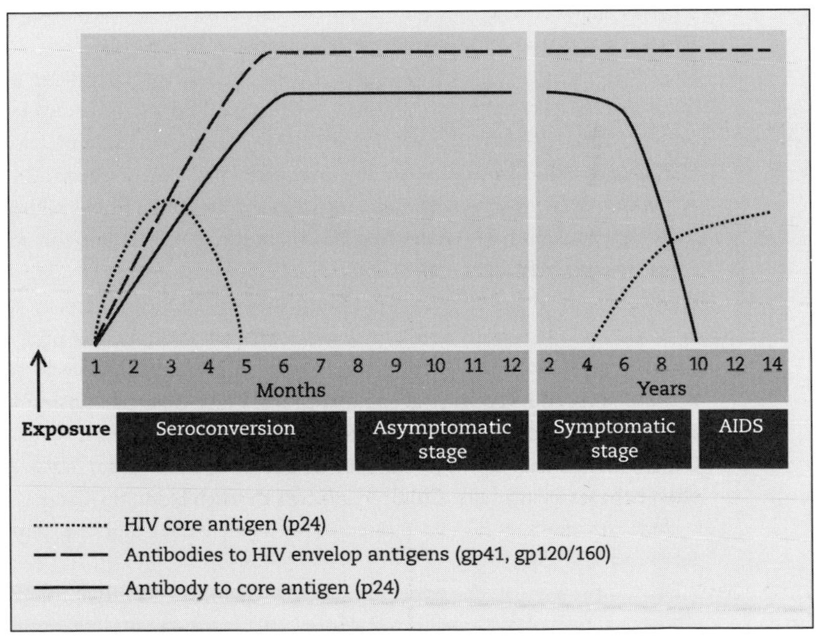

Fig 15.4 Typical immunological events during HIV infection.

• demonstration of p24 antigen (usually after dissociating maternal antibody)
• presence of viral nucleic material by PCR
• culture of virus from mononuclear cells.

Management

There is no cure for HIV/AIDS and the current therapeutic strategies are aimed at attempts to prolong the asymptomatic and symptomatic stages by using anti-retroviral drugs, prophylaxis of secondary infections and energetic treatment of such infections and cancers.

Anti-HIV drugs

The currently available antivirals, zidovudine, didanosine and zalcitaban are all nucleoside analogues which interfere with the reverse transcriptase enzyme of the virus. These agents probably modestly improve the survival of symptomatic patients (by about 1 year) but there are no reliable data to indicate their efficacy if used early during the asymptomatic stage. Emergence of resistant viruses within 1 year is probably an important factor. Combination therapies with two or more drugs are under evaluation and are being used in current clinical practice.

Prophylaxis against secondary infections

Start PCP prophylaxis in those with CD4+ $<200/mm^3$. Consider additional anti-*Toxoplasma* prophylaxis with trimethoprim if *Toxoplasma* serology is positive and the patient is receiving nebulised pentamidine as PCP prophylaxis. Consider rifabutin prophylaxis (for MAC) when CD4+ $<50/mm^3$. Vaccinate against pneumococci and hepatitis B. In children, monthly intravenous (IV) immunoglobulin injections will reduce the incidence of recurrent bacterial infections.

Prevention

• Health education highlighting the dangers of unprotected casual sex and the sharing of needles and syringes is of paramount importance. This should be general as well as targeted at high-risk groups
• The safety of donated blood, blood products and tissues must be ensured
• Health workers should be trained in the safe handling and disposal of sharps and needles
• Encouragement of condom use by the clients of commercial sex workers and sexually transmitted disease (STD) treatment centres
• The risk of materno-fetal HIV transmission may be reduced by the

avoidance of instrumentation during delivery, washing out of the birth canal, the use of zidovudine for the mother before and during delivery and for the baby after delivery, and delivery by caesarean section
- No effective HIV vaccine is currently available.

Prognosis

So far, the vast majority of HIV-infected persons have developed AIDS after 8–10 years of infection. Outside this group, a small number developed AIDS much more rapidly within 2–3 years (rapid progressors) whereas another small group (around 7 per cent) have remained well and immunocompetent even after 15 years from the time of infection (long survivors).

Length of survival after an AIDS-defining illness is highly variable, ranging from only a few months (disseminated MAC, cytomegalovirus (CMV) retinitis, lymphoma) to up to 4–5 years after Kaposi's sarcoma. Overall, median survival remains around 2 years.

FUNGAL INFECTIONS

Candidiasis

This is due to infection by the fungal yeast, *Candida*. Most infections are due to *C. albicans* which is a commensal of the human mouth and intestines. Disturbances to the body's intricate defence mechanisms underlie all infections but often remain unexplained. Less pathogenic candida like *C. tropicalis*, *C. glabrata* or *C. parapsilosis* may cause disease in severely immunocompromised patients.

Clinical features

These depend on the predisposing factors and sites of involvement (Table 15.3). The severity of mucocutaneous involvement depends on the degree of cell-mediated immunity (CMI) suppression, whereas systemic invasion usually occurs in neutropenic patients or in debilitated ill patients with venous access lines.

Treatment

Oropharyngeal candidiasis: mild cases respond to topical nystatin, miconazole or amphotericin B. Severely immunocompromised patients require an orally absorbable azole drug: fluconazole or itraconazole. Relapses are common and are best managed by intermittent self-therapy

FORMS OF CANDIDIASIS

Condition	Clinical features	Predisposing factors	Diagnosis
Oral			
Pseudo-membranous	White patches on buccal mucosa can be readily wiped off leaving a red, raw surface	Infancy, old age Debility, depressed CMI Steroid inhalation	Clinical Demonstration of *Candida* in scraping or smear if necessary
Acute atrophic	Sore mouth with reddened mucosa; tongue smooth, shiny, may be swollen	Oral broad-spectrum antibiotics	As above
Chronic atrophic	Chronic erythema and oedema of mucosa under the dentures; angular cheilitis may be present	Dentures Chronic mucocutaneous candidiasis*	As above
Oesophageal	Dysphagia May be silent	Severely depressed. CMI, as in AIDS	Endoscopy Barium-swallow (Fig. 15.5) Smear microscopy/culture
Vaginal	Vulvo-vaginal pruritus Local discomfort and thick curdy discharge White mucosal patches Vulval erythema Recurrences common	Common in apparently healthy Pregnancy Diabetes mellitus Tight insulating clothes Antibiotics Depressed CMI	Clinical, smear and culture
Penile	Soreness or irritation of glans penis Discharge, erythema or raised red patches	Intercourse with a person with vaginal candidiasis Diabetes	Clinical, smear and culture
Cutaneous	Erythema with tiny vesiculo-pustules Groin and	Warm, moist skin Friction and occlusion	Clinical, smear and culture

Continued

Table 15.3. Forms of candidiasis

FORMS OF CANDIDIASIS

Condition	Clinical features	Predisposing factors	Diagnosis
	submammary areas are commonly affected	Napkin dermatitis Chronic mucocutaneous candidiasis*	
Nail and nailfold	Swollen erythematous painful nailfold. Involvement of nail plate may cause discolouration, pitting, friability and separation Bacterial superinfection is common	Repeated immersion in water Chronic mucocutaneous candidiasis*	Clinical, smear and culture
Septicaemia	Fever, metastatic infections in liver, brain, heart, bones or joints may develop	Debilitated patients Indwelling venous lines IVDU using contaminated materials Neutropenia	Blood culture Tissue biopsy

* Chronic mucocutaneous candidiasis is a disease of unknown aetiology with congenitally impaired CMI to *Candida*.

Table 15.3. *Continued*

Vaginal candidiasis: single-dose fluconazole or a topical imidazole (clotrimazole, miconazole or econazole) for 5 days are effective

Cutaneous and nail infections: topical antifungal and steroid combination is useful. Chronic infection requires prolonged oral therapy with daily soaking of the nails using potassium permanganate or phenyl-mercuric borate solution

Systemic candidiasis: IV amphotericin B with or without oral flucytosine is the drug of choice. Infected lines must be removed.

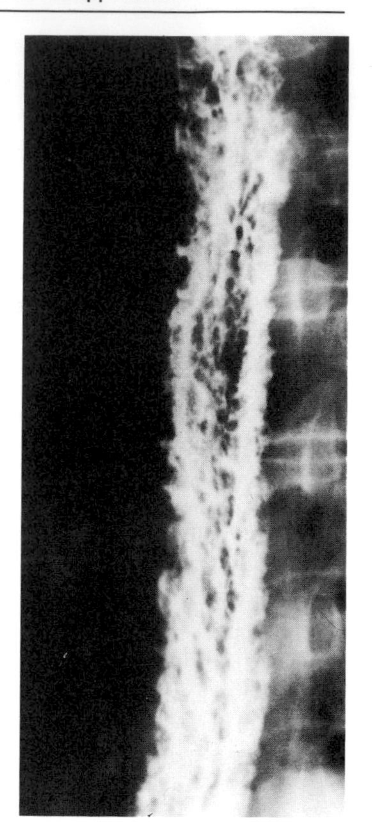

Fig 15.5 Oesophageal candidiasis on barium-swallow examination.

Aspergillosis

The causative organisms, *Aspergilla fumigatus*, *A. niger* and *A. flavus* are commonly found in soil, dust and decaying vegetable matter worldwide. Infection is via inhalation of airborne spores. Cases are usually sporadic but clusters of cases may occur in cancer wards from environmental sources.

Clinical features

Despite the ubiquity of the organisms, human illnesses are rare. Several forms are recognised.

Allergic bronchopulmonary aspergillosis

In atopic subjects colonisation and proliferation of *A. fumigatus* in the bronchial passages may lead to allergy to *Aspergilla*, causing cough and wheezy attacks. Mucosal damage may produce chronic obstructive airways disease and bronchiectasis. Transient lung infiltrates are visible on CXR due to bronchial plugging. Eosinophilic and serum precipitating antibodies are present. Specific IgE antibody to *Aspergilla* may be demonstrable. Treatment is as in asthma. Antifungal drugs are of little value.

Aspergilloma

A pre-existing lung cavity may become colonised leading to formation of a ball of fungal mycelium. Often asymptomatic, symptoms include cough and malaise; sometimes severe haemoptysis may develop. CXR shows a characteristic round shadow within a cavity with a halo round it. *Aspergilla* is found in sputum and serum precipitin test is positive.

Asymptomatic and mildly symptomatic patients are only observed, others may need surgical removal. Local instillation of antifungal drugs may help.

Invasive aspergillosis

In neutropenic immunocompromised patients *Aspergilla* may invade lung parenchyma, CNS, kidneys, bones and other organs. The condition has a high risk of fatality, but IV amphotericin B or itraconazole may be life-saving. If feasible, immunosuppressive therapy should be discontinued and neutropenia corrected.

In an outbreak situation, epidemiological investigations to locate the source of spores are necessary.

Other fungal infections of importance in immunocompromised patients

HISTOPLASMOSIS

Epidemiology

The causative organism *Histoplasma capsulatum* is widely found in soil in middle USA and Africa, and less commonly in the rest of the Americas and east Asia. It is rare in Europe. Infections occur in humans, animals and birds. Bird droppings and decaying organic matter in soil are prominent sources of infection.

Clinical features

- Infection occurs via inhalation of airborne spores and is usually asymptomatic. Less commonly, the patient develops an acute respiratory illness with cough, malaise and fever. CXR may show diffuse small opacities

- The disease is usually self-limiting and mild in the immunocompetent but, rarely, chronic disease resembling pulmonary tuberculosis develops
- Disseminated disease occurs very rarely in the immunocompetent but commonly in the immunocompromised such as AIDS patients. Presentation is subacute, with fever, weight loss, mild cough and splenomegaly. Miliary mottling may be present on CXR. Illness may be the result of reactivation of dormant infection.

Diagnosis

Serology is diagnostic in the immunocompetent but unreliable in immunocompromised patients. Demonstration of the organism in bone marrow, blood, sputum or splenic aspirate is necessary.

Treatment

Treatment is not necessary in the immunocompetent patient, who has usually recovered by the time of serological diagnosis. In disseminated disease, amphotericin B is the treatment of choice although fluconazole and itraconazole are also useful. Maintenance therapy is essential in AIDS patients.

COCCIDIOIDOMYCOSIS

This is caused by the fungus *Coccidioides immitus*. The organisms are only found in the dry desert areas of western USA, parts of south and central America. Transmission is via inhalation of airborne spores.

- Infection is usually asymptomatic or may result in a self-limiting, acute, influenza-like illness
- Disseminated disease may occur in the immunocompetent but is seen more commonly in the immunocompromised such as AIDS patients. Previously asymptomatic dormant infection may activate in such patients
- Amphotericin B is the drug of choice, although fluconazole is proving useful. Recurrences are common in disseminated infection and maintenance therapy is necessary.

CRYPTOCOCCOSIS

This is a systemic infection caused by the yeast-like fungus *Cryptococcus neoformans*, manifesting usually as meningitis.

Epidemiology

- Occurring worldwide, it was a rare infection in the pre-HIV era and was mostly seen in patients receiving immunosuppression therapy (lymphoma, sarcoidosis). Immunocompetent individuals normally have high resistance to clinical infections, but cases have occurred in patients without demonstrable immunodeficiency

• In immunosuppressed HIV patients, it is a major cause of life-threatening infection

• *Cryptococcus neoformans* is a saprophytic fungus present widely in soil and pigeon droppings and infections occurs through inhalation.

Clinical features

• Meningitis is the usual manifestation. Onset is insidious with worsening headache over many days. In AIDS patients the symptoms and signs of meningeal infection may be minimal or absent

• Widespread systemic dissemination may occur. Oral, pleural, pulmonary, myocardial, mediastinal, glandular and cutaneous cryptococcosis may be the presenting problem or the patient may present only with fever due to cryptococcaemia.

Diagnosis

A high index of suspicion is necessary and cryptococcal antigen should be looked for in serum in all HIV patients with pyrexia of unknown origin (PUO). CSF changes may be minimal with only a few lymphocytes, modestly raised protein and normal/low glucose levels. Cryptococcal antigen is usually positive in cryptococcal meningitis, but definitive diagnosis requires demonstration of the typical capsulated yeasts in centrifused CSF deposit stained by indian ink. Culture is confirmatory.

Treatment and prognosis

• IV amphotericin B with or without flucytosine for 6 weeks or longer is the treatment of choice although high-dose fluconazole may be equally effective. Relapse is common and maintenance therapy with fluconazole is necessary

• The disease has a significant mortality particularly in AIDS patients and the cure rate following either form of therapy is no more than 70%.

CYTOMEGALOVIRUS DISEASES

Epidemiology

A member of the herpesvirus group, CMV is found all over the world. Between 60 and 90 per cent of adults carry antibodies in their blood, signifying previous infection. Infection rate is higher in poorer socioeconomic groups and in sexually active individuals.

After primary infection, the virus persists in the body in a latent form.

Occasionally, reactivation occurs and the virus appears in the saliva, urine, cervical secretions, semen and breast milk, thereby facilitating transmission to others. Congenital infection may occur *in utero* or during birth.

Clinical features

Congenital infection

- Infection in babies of immune mothers is usually asymptomatic
- Infection *in utero* during the mother's primary infection carries a significant risk of:
  - symptomatic infection which may cause: still birth, retarded growth, jaundice, hepatosplenomegaly, purpura, encephalitis, microcephaly, choroidoretinitis, cerebral calcification
  - mental retardation which becomes apparent later but is asymptomatic at birth
- It is estimated that congenital CMV infection results in around 400 mentally handicapped children in the UK annually.

Acquired infection (in the immunocompetent)

Usually this is asymptomatic but rarely the following clinical syndromes may develop:

- pyrexial illness without specific features
- hepatitis with persisting fever. Liver histology may show granulomatous changes
- glandular fever-like illness with atypical mononucleosis but negative Paul–Bunnell test
- acute polyneuritis of Guillain–Barré type
- post-perfusion syndrome: a severe glandular fever-like illness which may follow transfusion of large quantities of infected blood, usually in the setting of open heart surgery.

CMV disease in the immunocompromised

This is due either to reactivation of latent infection or to primary infection in a seronegative recipient of blood or organ from a seropositive donor. The risk of clinical disease is greatest in the profoundly immunocompromised. Several clinical forms are recognised:

CMV pneumonitis: fever, non-productive cough and interstitial shadowing on the CXR. More common in transplant patients than in AIDS patients

Gastrointestinal CMV disease: mucosal ulceration may occur in the oesophagus, stomach, small intestine and colon causing dysphagia, haemorrhage or diarrhoea. Hepatitis and cholangitis may occur

CMV retinitis: this occurs most commonly in male homosexual HIV-positive patients with CD4+ <100/mm³. The condition usually begins unilaterally with blurred vision and field defects. The other eye is soon affected. The typical fundoscopic appearances are yellowish-white granular patches superimposed with flame-shaped haemomorrhages

CNS disease: subacute encephalitis, myelitis. Painful distal neuritis may occur.

Diagnosis

- In congenital infection, isolation of the virus in urine, presence of a compatible clinical picture and exclusion of other causes of similar presentation (herpes simplex virus, *Toxoplasma*, rubella) are necessary. Virus-specific IgM antibody is not routinely tested
- In older immunocompetent individuals, a four-fold rise in antibodies would indicate a primary infection
- In reactivation syndromes and in the immunocompromised, a definitive diagnosis requires demonstration of CMV inclusion in biopsy specimens. Isolation alone from body fluids has little diagnostic value
- CMV retinitis is a clinical diagnosis.

Treatment

- Two specific antiviral drugs are currently available: ganciclovir and foscarnet. Both have toxic side effects: bone-marrow suppression with ganciclovir and nephrotoxicity with foscarnet
- In CMV retinitis, either drug will stabilise or improve retinitis in around 80 per cent of patients. Maintenance therapy is essential to prevent relapses, which will occur eventually in most patients despite switching to the other drug. Blindness then occurs
- The efficacy of ganciclovir or foscarnet has been less impressive in gastrointestinal CMV disease, and disappointing in CMV pneumonitis.

Prevention

In view of the ubiquity of CMV in the community and its usual mode of transmission, avoidance of infection is not practicable. However, special measures are advisable in the following circumstances:

- pregnant women should be meticulous with hand washing if working with young children
- wherever possible, seronegative transplant patients should be given organs and blood from seronegative donors. Prophylaxis with hyperimmune gammaglobulin and ganciclovir or acyclovir will reduce the incidence and severity of primary infections in seronegative patients and of re-infections or reactivations in seropositive patients.

DISEASES CAUSED BY HUMAN T-CELL LEUKAEMIA/ LYMPHOMA VIRUSES

These are retroviruses and two types are known: HTLV-I and HTLV-2.

Epidemiology

• HTLV-I infection is endemic in Japan, the Caribbean Islands and parts of west Africa. HTLV-2 infection is prevalent among injecting drug users
• Transmission is through blood transfusion, injecting drug use, sexual contact and mother-to-child (breast feeding, *in utero*).

Clinical features

At the present time the role of HTLV-2 in disease is unclear, but HTLV-I causes two distinct clinical syndromes.

Adult T-cell leukaemia

• This is a rapidly progressive lymphoproliferative malignancy of mature T-cells. Peripheral blood changes of leukaemia, leukaemic skin infiltrations and hypercalcaemia are commonly associated and opportunistic infections occur due to immunosuppression
• People aged 20–60 years are mainly affected and there is usually a long incubation period of 20 years or more before the onset of the disease
• There is no effective treatment.

Tropical spastic paraparesis (HTLV-associated myelopathy)

• There is a slowly progressive demyelination of the spinal cord leading to symmetric upper motor neuron-type weakness of lower limbs, sphincter disturbances and sensory abnormalities. Magnetic resonance scans are useful in detecting cord lesions. Immune suppression does not occur
• Glucocorticosteroids may benefit some patients, but otherwise there is no treatment.

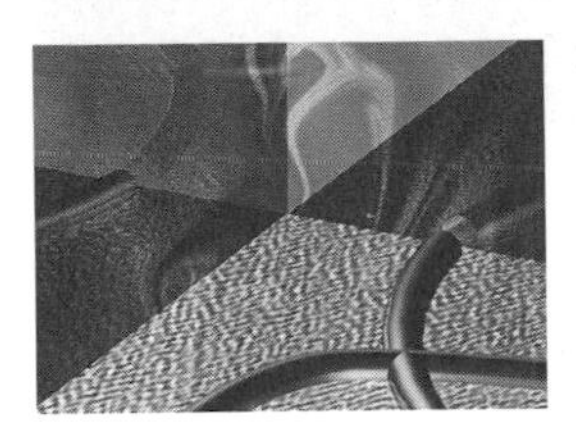

Congenital and Neonatal Infections

Some infections in pregnancy, during birth or in the neonatal period, can be much more severe than infections later in life. These have been described in other chapters, and this list with notes summarises the important congenital and neonatal infections.

INFECTIONS IN PREGNANCY

The following are infections that are important in pregnancy, because transplacental transmission may lead to fetal damage.

Rubella

Infection with rubella in the first trimester brings a high chance of congenital abnormalities (congenital rubella syndrome).

Toxoplasmosis

In early pregnancy this may cause a miscarriage; later, a stillbirth or a baby with chorioretinitis, intracerebral calcification and brain damage, jaundice and fever. Infection in later pregnancy is less harmful but may still cause chronic chorioretinitis.

Varicella

Varicella infection in early pregnancy may rarely lead to congenital varicella syndrome (hypoplasia of a limb and scarring in a dermatome).

Parvovirus B_{19}

This can cause fetal anaemia, fetal death and abortion. Congenital anomalies do not occur.

Syphilis

In pregnancy, this can result in low birth weight, deafness and various stigmata: interstitial keratitis, sabre shins and Hutchinson's teeth (congenital syphilis). Syphilis serological screening in the antenatal period, with penicillin treatment for the mother if positive, will reduce the risk.

Cytomegalovirus

Primary cytomegalovirus (CMV) infection during pregnancy is usually without symptoms in the mother but will infect 30–40 per cent of fetuses, producing clinically apparent disease in 10 per cent of the infected. The most severely infected babies have hepatosplenomegaly, intracerebral calcification, retinitis and pulmonary infiltration. Recurrent CMV in the mother rarely affects the fetus.

Malaria

This may cause a miscarriage or stillbirth. The risk outweighs the very low risk of chemoprophylaxis to pregnant women travelling to malarious areas. Congenital infection is rare.

INFECTIONS DURING LABOUR

Infections acquired at about the time of labour, with either intrauterine (amniotic) infection or infection during passage in the birth canal.

Varicella

Neonatal chickenpox is often much more severe when the contact is the mother (maternal rash within 1 week before delivery), or another person postnatally.

Herpes simplex

A primary attack of genital herpes in late pregnancy may cause generalised herpes infection in the neonate. This severe disease has a mortality of 70 per cent of cases.

Hepatitis B

Transmission to the baby usually occurs at the time of birth. Congenital infection leads to a long-term carrier state, with a risk of liver cancer.

Human immunodeficiency virus

The majority of mother-to-child transmissions occurs perinatally (during delivery, or less frequently via breast milk). Transplacental transmission is much less common.

Group B *Streptococcus*

This is a normal commensal of the human bowel and is an occasional commensal of the vagina. In infection during labour the neonate may have septicaemia and respiratory distress within hours of birth.

Listeria monocytogenes

Fetal infection may be due to materno-fetal transmission *in utero* or during passage through an infected birth canal. The infant may be still-born, born with septicaemia or develop neonatal meningitis. Mortality of neonatal infection is high.

INFECTIONS ACQUIRED IN THE NEONATAL PERIOD

- Newborn babies are colonised by whatever micro-organisms are in their environment and distributed by their carers
- Neonates, especially the premature, are susceptible to invasion by environmental microbes because they:
  - lack antibodies (except passively transferred maternal antibodies)
  - lack protective normal flora
  - often have thin skins, raw umbilical stumps, immature inflammatory response
  - may be compromised by openings offered by congenital malformations or medical interventions
- Thus skin infections, meningitis, gastroenteritis, epidemic viral diseases and septicaemia are prone to occur
- Outbreaks may occur due to *Staphylococcus aureus*, Gram-negative bacilli, coxsackie and echo viruses in neonatal nurseries.

SECTION 4

Zoonoses, Tropical Diseases and Helminths

CHAPTER 17

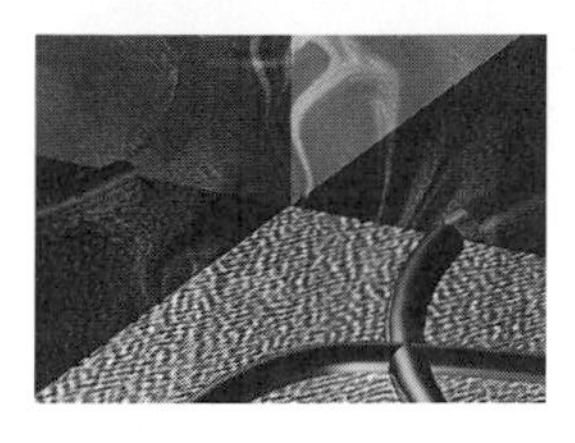

Non-helminthic Zoonoses and Tropical Infections

ZOONOSES

A zoonosis is an infection which is transmissible under natural conditions from vertebrate animal to humans. Zoonotic infections are common. Annually, from dogs alone, it is estimated that there are 31,000 episodes of wound sepsis due to *Pasteurella multocida*, 16,000 new infestations with *Toxocara*, 13,000 episodes of enteritis from *Campylobacter* infections and 9000 episodes of ringworm. Almost any animal infection can be transmitted to man if the conditions are suitable. The following are the most common and important. Zoonotic helminthic infections (*Toxocara*, hydatid disease, *Taenia*, *Trichinella*) are dealt with in Chapter 18.

Viral

- Rabies
- Orf, cowpox, lymphocytic choriomeningitis, Japanese B encephalitis, equine encephalitis, Kyasunur forest disease.

Bacterial

- *Campylobacter*, *Pasteurella*, *Salmonella*, *Leptospira*, *Brucella*, *Mycobacterium bovis*, *Coxiella burnetii*, *Chlamydia psittaci*, *Rochalimaea* (cat-scratch fever), *Borrelia burgdorferi* (Lyme disease)
- *Bacillus anthrax*, *Yersinia*, enterohaemorrhagic *Escherichia coli*, *Streptococcus suis*, *Francisella* (tularaemia), *Capnocytophaga* (DF 2), *Rickettsia*, *Listeria*, *Streptobacillus moniliformis* and *Spirillum minus* (rat-bite fever), *Clostridium tetani*.

Protozoal

- *Giardia, Cryptosporidium, Toxoplasma, Babesia*
- *Leishmania, Trypanosoma rhodesiense, T. cruzi.*

Fungal

- *Microsporum* spp., *Trichophyton* spp.

MALARIA

Epidemiology

Malaria is caused by parasites of the *Plasmodium* genus and is the most important protozoal infection worldwide.

- 300 million persons contract malaria annually and 2 million die, mostly children under 5 years of age; 10–15 deaths from *P. falciparum* occur annually in the UK
- *Plasmodium falciparum* cases in the UK have been increasing since 1970, resulting from increased travel, increasing resistance to prophylactic drugs, inadequate or neglected travel advice, and delayed diagnosis and treatment
- Four human species of malaria exist (*P. ovale*, *P. malariae*, *P. vivax* and *P. falciparum*)
- *Plasmodium vivax* predominates in India, Pakistan, Bangladesh, Sri Lanka and Central America, whereas *P. falciparum* is dominant in Africa and New Guinea; both are prevalent in South-East Asia, South America and Oceania. *Plasmodium ovale* and *P. malariae* occur mainly in Africa
- Transmission is through the bite of a female anophelene mosquito; man is the only reservoir
- The mosquito is initially infected by ingesting male and female gametocytes. A sexual cycle then takes place in the mosquito with sporozoites appearing in the salivary glands. These are injected during a feed
- Sporozoites quickly enter the liver (pre-erythrocytic phase) where they mature and multiply into schizonts which release numerous merozoites to invade red cells, where further maturation and multiplication take place (erythrocytic phase). The erythrocytic cycle lasts for 48 hours with *P. vivax* and *P. ovale*, and 72 hours with *P. malariae*, following which more merozoites are released which invade fresh cells. This gives rise to the typical tertian and quartan fever patterns. With *P. falciparum* it takes less than 48 hours, so that fever is daily
- In the case of *Plasmodium vivax* and *P. ovale*, the parasites may persist in the liver (hypnozoites), causing later relapses. In *P. falciparum* there are

no hypnozoites. *Plasmodium malariae* rarely relapses, due to persistent extra-hepatic parasites

• The average incubation period (IP) is approximately 12 days for *P. falciparum*, 13 days for *P. vivax*, 17 days for *P. ovale* and 28 days for *P. malariae* although infections may occur up to 3 months after leaving an endemic area for *P. falciparum*, 5 years for *P. vivax* and *P. ovale*, and 20 years for *P. malariae*.

Pathology and pathogenesis

Parasitised red blood cells have parasite-encoded proteins within their surface knobs which interact specifically with capillary endothelium. By sequestering in and obstructing the capillaries, parasitised cells cause anoxia, lactic acidosis, and leaky capillaries. This results in oedema, congestion and microhaemorrhage, which in turn lead to the complications of malaria. Many cytokines are increased during acute malaria (including tumour necrosis factor α (TNF-α)) but their exact role remains to be defined. Similarly, raised intracranial pressure has been linked with cerebral malaria but it is uncertain whether this contributes to coma or death. Immune complex formation (*Plasmodium* antigen, immunoglobulin G (IgG) and complement) in the kidneys of children with *P. malariae* may lead to the nephrotic syndrome. Tropical splenomegaly syndrome results from an unusual form of immune response to chronic malarial antigenaemia. Receptors on merozoites and on the red cells are essential for cell invasion. Sickle cell trait protects against severe *P. falciparum*, as does haemoglobin F; lack of the Duffy blood group antigen protects against *P. vivax*.

Clinical features

Malaria cannot be diagnosed with confidence clinically, nor can uncomplicated falciparum malaria be distinguished from the other three malarial species. Typically there is:

• an abrupt onset of fever and influenza-like symptoms

• periodicity developing after several days, with episodic fever separated by relatively symptom-free periods (unlikely in primary infections)

• episodes of fever consisting of cold, hot and sweating phases, in total lasting about 6–8 hours

• anaemia after several days

• hepatosplenomegaly in one-third and jaundice.

Complications

• Cerebral malaria, convulsions

• Severe normochromic anaemia

- Renal failure
- Pulmonary oedema (and adult respiratory distress syndrome (ARDS))
- Hypoglycaemia
- Shock state
- Spontaneous bleeding, disseminated intravascular coagulopathy (DIC)
- Haemoglobinuria (black water fever) (see Plate 7, between pp. 180 and 181)
- Recurrent infection with *P. vivax*, *P. ovale* and *P. malariae*
- Ruptured spleen (especially in *P. vivax*)
- Nephrotic syndrome with chronic *P. malariae* infection
- Tropical splenomegaly syndrome.

Several complications may occur simultaneously or sequentially. Cerebral malaria is strictly applicable when the patient is in an unrousable coma (defined by the Glasgow coma scale with exclusion of other encephalopathies), but is often applied when there is any degree of impaired consciousness and other sign of cerebral dysfunction. It usually develops after several days in adults but in children the history is often <2 days. Seizures, increasing obtundation and the development of abnormal neurology (e.g. dysconjugate gaze, decerebrate posturing) develop, sometimes with very rapid progression. Retinal haemorrhages are common. Coma may persist for several days after the clearance of parasites from the blood. Anaemia due to haemolysis and bone-marrow depression is common. Oliguric renal failure is a result of acute tubular necrosis, and peritoneal dialysis may be required. Acute pulmonary oedema may develop rapidly and often progresses to ARDS. Hypoglycaemia may result from high parasite counts or be a result of quinine-induced hyperinsulinaemia; it is more common in pregnant women and children. A shocked state (algid malaria) may develop and be related to concomitant septicaemia (usually Gram-negative). Thrombocytopenia is invariable and deranged clotting common in patients with high parasite counts. Haemoglobinuria is a result of intravascular haemolysis. Hyperpyrexia and acidosis may also occur and indicate complicated and severe disease.

Diagnosis

The differential diagnosis is wide, as *P. falciparum* malaria can mimic many diseases (e.g. hepatitis, meningitis, septicaemia, typhoid, arbovirus infection, typhus, gastroenteritis).

Nonspecific features supporting a diagnosis of malaria include:

- a consistent clinical and travel history
- splenomegaly
- thrombocytopenia.

Confirmation of malaria can be made by thick and thin film microscopy. Up to three smears should be examined if the history remains suggestive (see Plate 8, between pp. 180 and 181).

Treatment

Benign relapsing malaria (*P. vivax, P. ovale, P. malariae*)

• Chloroquine is the drug of choice for 3 days followed by
• Primaquine for 14 days (to eliminate the hepatic reservoir of infection and prevent relapses) in *P. vivax* and *P. ovale*
• In South-East Asia malaria, primaquine should be given for 3 weeks because of relative resistance
• Levels of the enzyme glucose-6-phosphate dehydrogenase (G-6-PD) must be checked before primaquine is given, since haemolysis may develop.

Malignant malaria (*P. falciparum*)

• Quinine for 3–7 days. Side effects are to be expected: tinnitus, deafness, nausea and vomiting (cinchonism)
• Intravenous (IV) quinine should be given if the patient:
  • is vomiting
  • has a high parasitaemia ($>$4 per cent)
  • has any complication as listed above
• Patients should be monitored for hypoglycaemia and cardiac arrythmias
• Quinine alone is not curative and fansidar or tetracycline also need to be given
• Other agents available for treatment are mefloquine, halofantrine and derivatives of qinghaosu (a Chinese herb)
• Intensive care is necessary in severe malaria
• Exchange transfusion efficiently reduces high parasite counts.

Prevention

• Vector control using insecticides and larvicides
• No vaccine has been convincingly shown to be protective, but trials are in progress
• Protection from biting is essential by use of knock-down insect sprays, mosquito bed-nets, insect repellants, mosquito coils, sensible clothing covering limbs and avoiding exposure at night
• Drug prophylaxis protects against developing disease. The choice depends upon the likelihood and type of malaria (countries, and areas within, being visited), the duration of intended stay, the prevalence of chloroquine-resistant *P. falciparum*, and host factors (age, pregnancy,

medical contraindications to certain drugs). The following are the most commonly used regimens in the major malarious areas:

- chloroquine (north Africa, the Middle East, China and Central America)
- mefloquine and standby therapy (South-East Asia)
- chloroquine and paludrine, or mefloquine (all other areas)

• Chemoprophylaxis should start 1 week before departing and continue for 4 weeks after returning. Chloroquine is the drug of choice for protection against *P. vivax*, *P. ovale* or *P. malariae*

• Chloroquine resistance is now a major problem in subsaharan Africa and South-East Asia

• Pregnant women, children < 10 kg, and persons requiring prophylaxis for > 12 months should not receive mefloquine; chloroquine and paludrine is the alternative

• Standby therapy is important for travellers to remote areas in malarious regions, especially in South-East Asia. Fansidar, mefloquine, quinine and halofantrine are the most important drugs in this category. Treatment for malaria should not be considered until at least 10 days have elapsed after entering a malarious area.

Prognosis

Mortality is exceptionally rare in *P. vivax*, *P. ovale* and *P. malariae* infection. For patients with complicated *P. falciparum* infection, mortality approaches 10–20 per cent. In those who survive cerebral malaria, about 5 per cent of adults and 10 per cent of children have neurological sequelae.

SOUTH AMERICAN TRYPANOSOMIASIS (CHAGAS' DISEASE)

Epidemiology

Chagas' disease is a zoonosis resulting from infection with *T. cruzi*. It occurs as both an acute and a chronic disease (affecting principally the heart) and is prevalent in Central and South America. Chagas' disease has the following characteristics:

• it affects over 20 million persons. Up to 10 per cent of suburban populations in Brazil are infected with the parasite

• it in its acute form, is a disease of the young, occurring before 10 years of age in 85 per cent. 25 per cent will go on to develop chronic disease at 35–45 years of age

- it is geographically variable in presentation (e.g. gastrointestinal tract involvement is uncommon in Venezuela)
- it is transmitted by reduviid bugs which feed off a wide range of mammalian species. They become infected by taking a blood meal from an infected host with circulating trypomastigotes. Transmission occurs when mucous membranes (most frequently the conjunctiva) or breached skin are contaminated by reduviid bug faeces which are produced during bug-feeding
- it is frequently transmitted by contaminated blood during transfusion (10,000–20,000 cases/year in Brazil alone). Rarely, transmission may occur *in utero* (3 per cent of infected mothers), during breast-feeding, from organ transplantation or nosocomially
- it may reactivate in immunosuppressed patients
- it has an IP of 1–4 weeks for acute, and 2–3 decades for chronic disease.

Pathology and pathogenesis

Following inoculation, pseudocysts develop which are collections of intracellular amastigotes, which then rupture releasing trypomastigotes into the blood. These are readily detectable during the acute stage. Lymphomononuclear infiltration is detectable around clustered parasites in the myocardium and myenteric plexus, where neuronal denervation occurs. In chronic disease, there is cardiac dilatation, apical aneurysm formation, mural thrombi and chronic lymphocytic infiltration; amastigotes are rarely found. Pathogenesis is incompletely understood. There is some suggestion that *T. cruzi*, heart muscle and nerve fibres share a common antigen which stimulates an autoimmune humoral and cellular response.

Clinical features

In vector-associated acute cases, lesions at the portal of entry are detectable in 50 per cent. These consist of:

- an inoculation chagoma (cellulitic lesion)
- Romana's sign (painless, unilateral, bipalpebral oedema of the eyelid with conjunctivitis)
- regional lymph-node enlargement.

Systemic manifestations are common, with fever, anorexia, malaise, myalgia and headache. On examination, there may be:

- hepatosplenomegaly and generalised lymphadenopathy
- oedema of the face and lower limbs
- pronounced tachycardia, indicating myocarditis.

Complications

- Meningoencephalitis
- Myocarditis
- Chronic disease (cardiomyopathy, megaoesophagus, megacolon).

Chronic cardiac disease occurs in 25 per cent of patients, but echocardiography (ECG) is abnormal in 60 per cent. Symptoms relating to cardiac failure (dyspnoea), arrythmias (dizziness) and thromboembolic episodes predominate. Dysphagia, reflux oesophagitis, pain and regurgitation of undigested food are the features of megaoesophagus, and constipation may suggest megacolon.

Diagnosis

In acute disease, diagnosis is suspected if:

- there is a history of exposure
- a chagoma or Romana's sign is present
- blood lymphocytosis is present.

Confirmation is by:

- examination of anticoagulated blood or buffy coat for motile trypomastigotes. Alternatively, stained thin and thick blood smears can be used
- serology, (positive 1–2 months after infection)
- xenodiagnosis, which involves using laboratory-reared reduviid bugs to feed off the patient's blood. These are then examined for parasites 30 days later
- mouse inoculation or blood culture on Novy–MacNeal–Nicolle (NNN) medium.

In chronic disease, the diagnosis is suspected if:

- there is a history of exposure
- there is evidence of cardiac disease, especially in a young adult and where there are conduction disturbances on ECG, a dilated cardiomyopathy or an apical aneurysm
- there is clinical and radiological evidence of dilated gastrointestinal tract disease.

Confirmation is by:

- serology
- xenodiagnosis
- mouse inoculation or blood culture.

Treatment

For acute disease, specific treatment must be prescribed because cure is possible. Two agents are available and given for 2 months:

- nifurtimox, or

- benznidazole.

For chronic disease, complications should be treated appropriately. Surgery has a role in gastrointestinal tract disease. Specific treatment has no place.

Prognosis

In acute disease, mortality is highest in congenital cases, immunosuppressed patients and young children. In areas of Brazil, cardiac disease is the most common cause of death in young adults. When disease results in heart failure, half will die within 2 years.

Prevention

- Vector control by insecticide spraying
- Housing improvement, health education, etc.
- Screening donated blood (or treating with gentian violet)
- No vaccine is available.

AFRICAN TRYPANOSOMIASIS

Epidemiology

African trypanosomiasis is a protozoal infection restricted to localised areas within Africa and caused by two epidemiologically distinct sub-species of *T. brucei*, *T. brucei gambiense* (west and central Africa) and *T. brucei rhodesiense* (east Africa). African trypanosomiasis:

- is transmitted by various species of tsetse flies. Their habitats are close to water (riverine – west African) or open country (savannah – east African)
- affects at least 20,000 cases annually; it is estimated that 50 million persons are at risk
- has principal reservoirs in humans (west African) and animals (bushbuck – east African). West African trypanosomiasis is likely in rural communities with river water contact, whereas east African trypanosomiasis is a sporadic disease of persons in contact with the animal reservoir, e.g. safari park visitors
- tends to be less severe, more chronic and associated with a lower-grade parasitaemia if due to *T. brucei gambiense*
- has an IP of from 2 to 28 days, being usually shorter in east African disease.

Pathology and pathogenesis

There are two stages, haemolymphatic (stage 1) and meningoencephalitic (stage 2). Following inoculation, there is a proliferation of trypanosomes with a lymphocytic inflammatory infiltrate. Trypanosomes then disseminate (stage 1), leading to systemic symptoms and reticuloendothelial hyperplasia. The characteristic cell is the morular cell of Mott (foamy plasma cell). Stage 2 is a diffuse meningoencephalitis with trypanosomes identifiable in the brain and spinal cord, associated with a lymphocytic and morular cell infiltrate. The cyclical fever pattern is a result of variations in the surface glycoprotein and allows the parasite to evade host immune responses more effectively.

Clinical features

In the acute stages of illness

- An indurated, painful chancre at the bite site usually develops (much more frequently with east African trypanosomiasis); regional lymphadenopathy often occurs
- Fever and influenza-like symptoms are typical. The fever may develop into a pattern of remission and relapse, each phase lasting approximately 1 week
- A transient macular rash may be seen
- Hepatosplenomegaly and mobile rubbery lymphadenopathy develop, especially suboccipitally (Winterbottom's sign – more common in *T. brucei gambiense*).

In the chronic stages of illness

- Central nervous system (CNS) involvement is common and characterised by intractable headache, choreoathetoid movements, ataxia, behavioural abnormalities, cranial nerve lesions, deep hyperaesthesia and later drowsiness, confusion and somnolence (sleeping sickness)
- Endocrine dysfunction, weight loss and anaemia may occur.

Complications

- Myocarditis, pericardial effusion
- Pulmonary oedema
- Acute fulminant presentation (mimicking severe falciparum malaria).

Diagnosis

This is made by:

- wet preparations or thick and thin Giemsa stains of peripheral blood or buffy cells (see Plate 9, between pp. 180 and 181)
- stained aspirates from chancres, lymph-nodes, bone marrow or cerebrospinal fluid (CSF)

- serology. This is useful in non-indigenous persons and in seroprevalence surveys, but not in indigenous persons where it may reflect past exposure
- CSF examination. This is essential in all patients with trypanosomiasis, as it determines treatment. Typically in stage 2 CNS disease, there is a raised pressure, $>5 \times 10^6$ lymphocytes/litre, raised total protein, raised total IgM and morular cells
- culture and animal inoculation.

Treatment

The choice of drug depends on the presence or absence of CNS involvement.

- Suramin and melarsoprol are effective against both subspecies, but suramin is the drug of choice for stage I disease because of melarsoprol's toxicity. However, suramin is not effective for CNS disease, where melarsoprol is the drug of choice
- Pentamidine and difluoromethylornithine are alternatives for *T. brucei gambiense*; difluoromethylornithine penetrates the CNS
- It is imperative to start treatment (usually with suramin) before doing a lumbar puncture (LP) because of the risk of introducing trypanosomes into the CNS
- Corticosteroids may reduce the development of reactive encephalopathy during melarsoprol therapy.

Prevention

This is by:

- effective surveillance and treatment schedules
- vector control with insecticides and avoiding bites
- appropriate land use.

Prognosis

Stage I disease can be cured, although fulminant infection as seen in non-indigenous travellers has a significant mortality. Stage 2 disease has a greater morbidity and mortality, with less effective and more toxic treatment. Reactive encephalopathy induced by melarsoprol occurs in 5 per cent and has a mortality rate of 50 per cent.

LEISHMANIASIS

Leishmaniasis is caused by the protozoan genus *Leishmania*, and can be divided into visceral (kala-azar), cutaneous and mucocutaneous syndromes. The four important species for humans are *L. donovani*, *L. tropica*, *L. mexicana* and *L. braziliensis*. They are transmitted to humans by sandflies and the life cycle involves amastigote (human) and promastigote (sandfly) forms. They are intracellular pathogens. With the exception of *L. donovani* and *L. tropica* in India, the disease is a zoonosis with the major reservoirs being rodents and dogs.

Old-world cutaneous leishmaniasis

Epidemiology

Leishmania tropica, *L. major* and *L. aethiopica* are the causes of a self-limiting, indolent ulcer affecting exposed sites. This form of cutaneous leishmaniasis:

- is found in the Mediterranean, the Indian subcontinent, China and subsaharan Africa
- embraces three subspecies; *L. major* (rural, desert rodent reservoir, multiple ulcers), *L. tropica* (urban, human and dog reservoir, single ulcer) and *L. aethiopica* (rural, rock hyrax reservoir, non-ulcerative)
- is influenced by the host's immune response
- is a sporadic disease with occasional outbreaks during road construction, etc.
- usually has an IP of 2–4 weeks (rarely up to 3 years).

Pathology and pathogenesis

Amastigotes within macrophages at the bite site elicit a granulomatous inflammatory reaction. Eradication and immunity occur over 6–12 months and the patient develops cutaneous hypersensitivity. Where immunity fails to develop, the organisms disseminate through the skin without ulceration (diffuse cutaneous leishmaniasis). Where the immune response is excessive, a papular reaction over and around the scar develops (leishmania recidivans).

Clinical features

A painless papule forms 2–4 weeks after the bite, which enlarges over 3 months into a circular ulcer with raised, indurated edges and satellite lesions. Most heal spontaneously over 1 year, leaving a flat, depigmented atrophic scar. Regional lymphadenitis is common.

Complications

- Diffuse cutaneous leishmaniasis
- Leishmania recidivans.

Diagnosis

Differential diagnosis of the ulcerated form includes an infected mosquito bite, tropical ulcer, cutaneous diphtheria, *Mycobacterium ulcerans*, sporotrichosis and amoebic ulcer. Differential diagnosis of the papular form includes lepromatous leprosy, tuberculosis and sarcoidosis.

- Specimens are collected by split-skin smears, curettings or by biopsy of the ulcer edges
- Leishmaniasis is confirmed by identification of the organism by Giemsa smear, or culture on NNN medium
- The leishmanial skin test is helpful in non-indigenous persons and becomes positive after 2 months.

Treatment

- No treatment is usually needed
- For cosmetically unsightly or large multiple lesions, local therapy with paromomycin cream, sodium stibogluconate injections or cryotherapy can be used
- Rarely, parenteral agents are indicated (e.g. for diffuse cutaneous leishmaniasis) (see p. 291).

Prevention

- Avoidance of being bitten (sensible clothing, insect repellant, building dwellings away from forest, bed-nets)
- Insecticide sprays in houses; control of sandfly breeding
- Control of the animal reservoirs (dog licensing and destroying strays; rodent control).

Prognosis

Old-world cutaneous leishmaniasis is a benign disease. Both diffuse cutaneous leishmaniasis and leishmania recidivans are chronic illnesses lasting 20 or more years.

Visceral leishmaniasis

Epidemiology

Visceral leishmaniasis is a chronic systemic disease caused by *L. donovani* (India and east Africa), *L. infantum* (Mediterannean) and *L. chagasi* (Central and South America). It is characterised by fever, hepatosplenomegaly and pancytopenia.

- The reservoirs are dogs (Central and South America, Mediterranean, China, Asia), rodents (east Africa) and humans (India)

- Disease may be sporadic (non-indigenous, adults), endemic (indigenous, children <5 years of age) or epidemic (indigenous, all ages)
- Asymptomatic infections are common
- It affects children and young adults
- The IP is 2–4 months.

Pathology and pathogenesis

Following inoculation, amastigotes disseminate to the reticuloendothelial system where they result in enlargement (spleen, liver, lymph-node) or replacement (bone marrow). Polyclonal humoral stimulation occurs, resulting in hyperglobulinaemia but with defective cell-mediated response and an inability of T-cells to activate macrophages to kill the parasite.

Clinical features

In the classic form of the disease:

- there is an insidious onset with fever (twice-daily peak), sweating and malaise. Weight loss, diarrhoea, a dry cough and epistaxis then develop
- hepatosplenomegaly, pancytopenia and wasting develop over 1–2 months; the spleen may become enormous. Lymphadenopathy may be present
- with progressive emaciation, the skin may darken, hair fall, and petechiae and bruising develop.

Complications

- Intercurrent infection (e.g. pneumonia, tuberculosis)
- Internal haemorrhage
- Renal amyloidosis
- Post kala-azar dermal leishmaniasis
- Mucosal spread.

1–3 per cent (Africa) and 10–20 per cent (India) of patients develop post kala-azar dermal leishmaniasis 1–5 years after treatment, restricted to the skin (especially the face) with progressively hypopigmented patches, butterfly erythema amd multiple nodules developing.

Diagnosis

The differential diagnosis is wide and includes typhoid, brucellosis, lymphoma, disseminated tuberculosis, tropical splenomegaly syndrome, hepatic schistosomiasis and lepromatous leprosy.

Nonspecific features supporting a diagnosis of visceral leishmaniasis are:

- normocytic normochromic anaemia, leucopenia and thrombocytopenia

• raised IgG, total globulin and erythrocyte sedimentation rate (ESR) with a low albumin.

Confirmation can be made by:

• demonstration of amastigotes (Leishman–Donovan bodies) from lymph-node (60 per cent), bone-marrow (85 per cent) or splenic (98 per cent) aspirate, liver biopsy (60 per cent) or buffy coat cells
• culture on NNN medium
• serology.

Treatment

Parenteral therapy is necessary.

• Stibogluconate or amphotericin B are the standard treatments
• Pentamidine or paromomycin are alternative therapies
• Paromomycin and stibogluconate in combination is better than either alone
• Allopurinol in combination with pentavalent antimonials may also be effective.

Prevention

• Avoidance of being bitten
• Insecticide sprays in houses; control of sandfly breeding
• Control of the animal reservoir
• Control of the human reservoir by prompt treatment of human cases and case-finding
• Health education (covering of cutaneous lesions).

Prognosis

Prognosis is good but, without treatment, nearly all patients die within 2 years. Follow-up is imperative to detect relapses which occur in up to 25 per cent of cases.

New-world cutaneous leishmaniasis

Epidemiology

New-world leishmaniasis causes single or multiple ulcers (identical to old-world *L. tropica* infection) and mucocutaneous disease.

• Two species account for the benign, self-limiting ulcers, *L. mexicana* and *L. peruviana*. *Leishmania braziliensis* infections often result in mucocutaneous disease
• The primary reservoirs are the dog (*L. peruviana*) and rodents (*L. mexicana*, *L. braziliensis*); domestic animals may serve as secondary reservoirs

• The IP is 2–8 weeks for localised cutaneous disease and 1 month to ≥1 year for mucocutaneous disease.

Pathology and pathogenesis

The pathogenesis of localised cutaneous ulcers and diffuse cutaneous leishmaniasis is the same as for *L. tropica* infections (see p. 288). Mucocutaneous disease is associated with scarce amastigotes and prominent mononuclear cell infiltrate.

Clinical features

Mucocutaneous disease is characterised by the reappearance of mucosal lesions months to years after the disappearance of the primary ulcer. The mucosal disease results in a mutilating and destructive process affecting the nose, oral cavity and pharynx.

Complications

- Aspiration pneumonia
- Airways obstruction.

Diagnosis

The differential diagnosis is wide and includes lepromatous leprosy, South American blastomycosis, yaws, neoplasms, sporotrichosis and tertiary syphilis.

Nonspecific features supporting a diagnosis of new-world leishmaniasis are:

- history of residence in an endemic area
- classical destructive features of mucocutaneous disease.

Confirmation of new-world leishmaniasis is through:

- identification of amastigotes
- isolation of *Leishmania* on NNN medium
- positive serology (60–95 per cent), although there is some cross-reaction with Chagas' disease
- a positive *Leishmania* skin test (85–100 per cent).

Treatment

- In areas where mucocutaneous disease is prevalent, all skin ulcers should be treated, even if they appear to be uncomplicated
- Pentavalent antimonial compounds are the drugs of choice (sodium stibogluconate, pentamidine)
- Allopurinol alone or in combination with pentavalent antimonials appears more effective than pentavalent antimonial monotherapy
- Amphotericin B is an alternative agent

- There is a considerable relapse rate with established mucocutaneous disease (50 per cent of cases)
- Plastic surgery may be necessary.

Prevention

- Because the animal reservoir is wild, little can be done to control this element
- Avoidance of being bitten
- Insecticide sprays in houses; control of sandfly breeding.

Prognosis

In its worst form, this is an inexorably progressive and mutilating condition which is occasionally fatal.

TOXOPLASMOSIS

Epidemiology

Toxoplasma gondii is a zoonosis of worldwide distribution which causes a mild or subclinical illness in healthy persons. However, in patients with immunodeficiency (e.g. acquired immunodeficiency syndrome (AIDS)), or in children infected *in utero*, the disease can be severe. *Toxoplasma gondii:*

- has a definitive host in cats, where the organism's sexual cycle takes place. Faecal excretion of oocysts occurs in <1 per cent; these are ingested by many animals (e.g. mice and birds), where the asexual cycle takes place in tissues, which are subsequently eaten by the cat. Humans are infected either by ingesting oocysyts (common) or by eating undercooked meat (uncommon)
- is acquired very rarely at transplantation (particularly cardiac), by blood transfusion or as an occupational hazard of laboratory workers
- seropositivity increases with age (UK prevalence of 8 per cent in children <10 years of age rising to 47 per cent in those >60 years), with a seroconversion rate of 0.5–1 per cent/year
- infects 0.3 per cent of pregnant women in the UK during pregnancy, in 30 per cent of which the fetus is infected. *In utero* transmission is lowest in the first trimester (15 per cent) and highest in the third (60 per cent); severity is greatest when infection occurs in the first trimester. However, congenital toxoplasmosis is uncommon: less than 20 cases are reported each year in England and Wales
- causes subclinical infection in 80–90 per cent. It is estimated to cause 5 per cent of cases of clinically significant lymphadenopathy

• tissue cysts persist for the life of the host and are the source of recrudescence
• presents at birth (congenital), or as a young adult (lymphadenitis (acquired) or choroidoretinitis (congenital))
• has an IP of 5–23 days.

Pathology and pathogenesis

Following entry, the parasites disseminate widely via blood and lymphatics; any tissue or organ may be invaded. The infection is checked by cytokine-induced macrophage-mediated destruction; later, humoral immunity is important in the induction of the cyst formation. *Toxoplasma* trophozoites have the ability to prevent lysosome–phagosome fusion and acidification of the lysosome within macrophages. The encysted *T. gondii* parasites remain viable until the death of the host. The characteristic feature on lymph-node biopsy is a stellate abscess. In congenital infection, tissue necrosis results in microcephaly, choroidoretinitis and intracranial calcification.

Clinical features

Four main clinical syndromes result from human infection with *T. gondii*: lymphadenitis, choroidoretinitis, congenital infection and cerebral toxoplasmosis (human immunodeficiency virus (HIV)-positive patients). Toxoplasmal lymphadenitis presents with:

• lymphadenopathy, mainly cervical and usually non-suppurative and non-tender
• an influenza-like illness with low-grade fever
• occasional hepatosplenomegaly.

The symptoms and signs may wax and wane, but resolve within a few months.

Complications

• Choroidoretinitis
• Myocarditis, pneumonitis, encephalitis, hepatitis
• Congenital infection
• Chronic fatigue syndrome.

Ocular toxoplasmosis usually results from congenital disease but can follow on from acquired infection. It is a focal necrotising retinitis which heals with clearly demarcated edges enclosing pale atrophic retina with black pigment. Symptoms include blurred vision, pain and photophobia. *Toxoplasma* acquired during pregnancy may lead to spontaneous abortion, stillbirth and premature delivery. The majority of infected live offspring do not have detectable disease at birth but do carry a significant risk of developing ocular or neurological disease in later life.

Those with overt clinical disease as neonates tend to progress to severe sequelae.

Diagnosis

Nonspecific features supporting a diagnosis of toxoplasmosis are:

- history of contact with cats or ingestion of raw meat
- heterophile antibody-negative glandular fever-like illness
- typical histology on lymph-node biopsy.

Confirmation is by:

- serology. The tests used are the dye, Latex agglutination and haemagglutination (measuring IgG antibody), and specific IgM and IgA antibody detection assays. Either a four-fold rise or fall between acute and convalescent sera in IgG or the presence of IgM or IgA antibody is needed to confirm recent infection
- demonstration of *Toxoplasma* trophozoites in histological specimens
- isolation of *T. gondii* (difficult)
- DNA amplification (polymerase chain reaction (PCR)).

Treatment

- No therapy is required for lymphadenitis
- For the immunocompromised with cerebral or other systemic site for disease, and for ocular disease:
  - pyrimethamine, with sulphadiazine or clindamycin
- For congenital toxoplasmosis:
  - pyrimethamine and sulphadiazine, or spiramycin.
- For pregnant women recently infected:
  - spiramycin (throughout pregnancy)
  - pyrimethamine and sulphadiazine (after first trimester)
- Steroids may be indicated in choroidoretinitis
- Folinic acid should be given to prevent the antifolate effect of pyrimethamine.

Prevention

- Primary and secondary prophylaxis in HIV persons with CD4+ counts $<200/mm^3$: co-trimoxazole is the most suitable agent
- Identification and treatment of infection during pregnancy
- Treating meat thoroughly to render cysts non-infectious
- Avoidance of cat faeces for those at risk (immunosuppressed, pregnant women).

Prognosis

Acquired toxoplasmosis in an immunocompetent patient is a benign illness. Ocular disease is a rare cause of blindness. Cerebral toxoplasmosis is a severe disease with a 10 per cent fatality and 10 per cent incidence of severe neurological complications.

Cross-references

Cerebral toxoplasmosis (see p. 254).
Congenital toxoplasmosis (see p. 271).

RABIES

Epidemiology

Rabies is a viral zoonosis causing a fatal encephalomyelitis. It is estimated that there are >50,000 deaths per year worldwide, nearly all occurring in developing countries. Rabies:

- is caused by a single-stranded RNA rhabdovirus
- is endemic, except in the UK, Australia, Antarctica, Japan, Scandinavia, Portugal, Greece and many small islands
- infectivity in animals (except vampire bat) is from a few days before they become unwell to death (usually within 2 weeks). Very occasionally an asymptomatic dog may excrete the virus. Bats are healthy carriers
- is maintained in a wild animal reservoir which varies according to country: e.g. fox in Europe, Canada and Arctic regions; skunk, raccoon and bat in North America; wolf in western Asia; mongoose and jackal in Africa and vampire bat in Central and South America. Occasionally, rabies is contracted directly from the wild animal
- strains do not spread readily through heterologous species (e.g. between dogs and foxes)
- can affect most wild animals, but susceptibility varies according to species (e.g. wolves susceptible, opossums resistant)
- in animals presents in two main forms: furious in dogs (excitation and aggression) and dumb in foxes (lethargy and uncharacteristic tameness), death occurring in 4–8 days.

Rabies in humans:

- results from a dog (82 per cent) or cat (10 per cent) bite or scratch with virus-laden saliva: very rarely transmission may occur through mucous membranes, aerosol (bats in infested caves), or implantation (corneal transplants). The virus cannot penetrate intact skin
- is more likely to be acquired and have a shorter IP if (i) the site of the bite is the face; (ii) the victim is young; (iii) the injuries are severe or multiple; (iv) the bite was not through clothing; (v) the animal was a wolf; and (vi) no pre- or post-exposure prophylaxis was given
- causes an encephalitis where mortality approaches 100 per cent; there

have been three reported recoveries in persons who had received pre- or post-exposure prophylaxis
• is especially common in India (25,000–50,000/year), China (6000–7000/year) and Bangladesh (2000/year)
• has an IP of 2–8 weeks (range 10 days–2 years).

Pathology and pathogenesis

Initial amplification of virus occurs in muscle around the bite site, followed by attachment via the glycoprotein to nerve endings (probably acetylcholine receptors) and retrograde axonal transport along peripheral nerves (3 mm/hour). The characteristic histological feature is the Negri body (a viral inclusion body). In later stages, virus spreads centrifugally by peripheral nerves to most tissues, including the salivary glands, where it is shed in the saliva.

Clinical features

Rabies may be described as 'furious' or 'paralytic', depending on whether the brain or spinal cord respectively are predominantly affected. Initial features for both are:
• discomfort, itching or paraesthesiae at the site of the healed bite wound
• anxiety, agitation
• fever, headache, myalgia, sore throat.

Furious rabies is characterised by:
• hydrophobia (fear of drinking which brings on inspiratory muscle spasms), aerophobia (spasm precipitated by cold air) and excessive salivation (see Plate 10, between pp. 180 and 181)
• marked agitation but with lucid intervals
• meningism, cranial nerve and upper motor neuron signs
• autonomic stimulation.

Paralytic rabies is characterised by:
• an ascending asymmetrical or symmetrical paralysis.

Complications

• Convulsions, opisthotonos
• Cardiac arrythmias, renal failure
• Cerebral oedema, hyperpyrexia, hypothalamic disturbance
• Development of flaccid paralysis
• Coma, respiratory or cardiac arrest.

Diagnosis

The disease must be distinguished from hysteria, encephalitis (both primary and following rabies vaccine), tetanus, bulbar poliomyelitis, and Guillain–Barré syndrome. Nonspecific features supporting a diagnosis of rabies include:

- the presence of hydrophobia (50 per cent of cases)
- lymphocytic CSF with normal/slightly raised protein
- history of a dog bite acquired in a rabies-endemic country
- the absence of muscle rigidity between spasms.

Confirmation of rabies in the patient can be made by the:

- demonstration of rabies antigen by antibody techniques in corneal smear, skin biopsy from the back of the neck at the hairline or from brain biopsy
- isolation of rabies virus from saliva by mouse inoculation or tissue cell culture (takes 7 days)
- demonstration of a rising antibody titre in serum and CSF
- demonstration of Negri bodies on brain biopsy.

Confirmation of rabies in the animal can be made by the demonstration of rabies antigen, Negri bodies on brain biopsy, or culture of rabies virus.

Treatment

Where there is no access to intensive care facilities, patients should be given adequate opiate analgesia to relieve terror and pain. Intensive care is the only method of prolonging (or possibly saving) life and is directed at preventing complications. The patient should be strictly isolated and the staff wear goggles, masks and gloves in addition to being immunised.

Prevention

- Import controls and quarantine of imported animals; control of stray dogs, etc.
- Regular immunisation of dogs
- Immunisation of wild animal reservoir using baits and attenuated vaccine
- Pre-exposure immunisation of high-risk groups (veterinarians, wildlife conservation personnel, etc.)
- Early treatment of wounds with soap and water and virucidal antiseptic (e.g. povidone iodine, alcohol)
- Post-exposure active immunisation with human diploid cell vaccine (HDCV) and passive immunisation with hyperimmune rabies immunoglobulin (RIG).

Where there has been a minor exposure (licks on skin, scratches or abrasions, minor bites through clothes on extremities):

- give HDCV on days 0, 3, 7, 14, 30 and 90

- observe the animal and stop treatment if animal is healthy after 10 days post-bite
- administer RIG if rabies confirmed in the animal. Half of the total dose is given intramuscularly and half infiltrated into the wound.

Where there has been a major exposure (licks of mucosa, major or multiple bites):

- give both vaccine and RIG
- observe the animal and stop treatment if animal is healthy after 10 days post-bite.

Prognosis

Clinical rabies is almost always fatal. Without intensive care, one-third of victims die early during a hydrophobic spasm. The remainder lapse into coma and develop generalised flaccid paralysis.

TROPICAL VIRAL INFECTIONS

Viral haemorrhagic fevers (VHFs) are a group of distinct acute viral infections which cause various degrees of haemorrhage, shock and sometimes death. Many are rarely identified and most are geographically very restricted. This section deals with the major VHFs: yellow fever, dengue, Lassa fever and hantaviruses. There are certain common features:

- non-haemorrhagic disease is common and death from blood loss is rare
- most viruses have a zoonotic reservoir (e.g. Lassa fever: the rat)
- many have an insect vector (e.g. Congo–Crimean haemorrhagic fever: ticks)
- most infections occur in remote areas, leading to under-reporting
- some (Lassa, Ebola, Marburg and Congo–Crimean) can be transmitted from person to person, when morbidity and mortality tend to be greater. Lassa fever is the main VHF with this potential. Ebola had not been reported since 1976 but in 1995 caused an outbreak in Zaire. Reports of Marburg are very infrequent. Congo–Crimean haemorrhagic fever is endemic in the Middle East and eastern Europe, but the risk of transmission is much lower.

Lassa fever

Epidemiology

Lassa fever:

- is endemic in west Africa; Sierra Leone, Liberia and Nigeria are the

main endemic foci. Up to 50 per cent of the indigenous population have serological evidence of past infection
• results in 300,000 cases and 5000 deaths annually
• is spread mainly by contact with aerosolised urine from chronically infected multimammate rats
• must be suspected in any febrile patient recently returned from an endemic area
• has a maximum IP of 18 days, being shorter for secondary cases.

Pathology and pathogenesis

Lassa virus is a single-stranded arenavirus with a genome coding for a viral polymerase and several structural proteins. With the development of immunity, antibodies against surface glycoproteins develop but these are not very effective at neutralising the virus. Following acquisition, a viraemia develops with, in severe cases, the development of a leaky capillary syndrome.

Clinical features

Typically there is:
• gradual onset of influenza-like symptoms
• exudative pharyngitis, conjunctivitis and proteinuria.

Complications

• Shock and leaky capillaries (causing ARDS, pleural effusions and oedema of the face and neck)
• Encephalopathy
• Pericarditis and congestive cardiac failure
• Sensorineural deafness.

Diagnosis

Features consistent with a diagnosis of Lassa fever are:
• travel to a rural endemic area within the last 21 days
• potential exposure to reservoir (rats or working as a health-care worker)
• pharyngitis, conjunctivitis and bleeding
• negative malarial films
• early lymphopenia followed by a neutrophilia.

A restricted list of laboratory specimens, namely blood and faeces for culture (typhoid), further blood films (malaria), and midstream urine (MSU) (pyelonephritis) are collected and examined in secure laboratories.

Confirmation is by:

- serology, with a four-fold rise in IgG antibodies on acute and convalescent sera, or the detection of IgM antibodies
- viral culture.

Treatment

- Strict isolation. If suspicion is high, the patient should be cared for in a high-security isolation unit with full-length gowns, gloves, masks, overshoes and caps
- IV ribavirin reduces mortality if given before the 7th day (from 61 per cent to 5 per cent) in severe cases.

Prevention

- Surveillance of close contacts
- Ribavirin to close contacts.

Prognosis

The overall mortality is 2 per cent. For hospitalised cases this increases to 10–15 per cent. There is a higher mortality in pregnancy (30 per cent). High transaminases are predictive of a poor outcome.

Yellow fever

Epidemiology

Yellow fever is due to a flavivirus and is mosquito-borne (principally *Aedes aegypti*). Recent epidemics in Nigeria have caused more than 150,000 cases and 30,000 deaths. Yellow fever has the following characteristics:

- it is a zoonosis of monkeys endemic in Africa and South America
- it may cause epidemics in two forms: 'jungle' (when humans are clearing forests with mosquitos breeding in tree-holes) and 'urban' (densely populated areas with mosquitos breeding in man-made containers)
- it has an IP of 3–6 days.

Pathology and pathogenesis

Following inoculation, the virus replicates in draining lymph-nodes and then causes viraemia. A characteristic midzonal hepatic necrosis occurs. The kidneys show evidence of acute tubular necrosis.

Clinical features

The illness ranges from mild to catastrophic. Typically there is:

- an acute biphasic fever with abrupt onset of severe influenza-like illness, conjunctivitis and relative bradycardia
- after remission for a few days, toxicity returns with severe vomiting, abdominal pain, jaundice, renal failure, hypotension and haemorrhage.

Complications

- Shock, coma and convulsions
- Myocarditis
- Hepatitis.

Diagnosis

Yellow fever has to be distinguished from falciparum malaria, leptospirosis, fulminant viral hepatitis, dengue haemorrhagic fever (DHF) and other VHFs. In favour of a diagnosis of yellow fever are the:

- absence of a history of immunisation against yellow fever
- history of visiting an endemic area
- biphasic illness with hepatitis, DIC and renal failure
- thrombocytopenia and leucopenia.

Confirmation is from:

- viral antigen detection by enzyme-linked immunoadsorbent assay (ELISA)
- viral culture
- detection of IgM antibodies
- four-fold rise in IgG antibodies between acute and convalescent sera.

Treatment

Supportive treatment in intensive care is necessary for severe cases.

Prevention

Excellent immunity can be achieved with a live attenuated vaccine (Yellow Fever 17D) which has a long-lasting immunity (10 years at least) and is extremely safe. A vaccination certificate is needed for international travel to and from most countries with endemic yellow fever.

Prognosis

When jaundice is present, 20–60 per cent of patients die. Death may be a result of fulminant hepatitis, renal failure, myocarditis or DIC.

Dengue fever

Epidemiology

Dengue is the most important of the arboviruses, with 40–80 million people becoming infected worldwide each year. Characteristics are as follows:

- it is endemic throughout the tropics; there were 106 cases imported into the UK in the first 9 months of 1990
- it is transmitted by *A. aegypti*
- it is present as four serotypes, each of which can infect humans. In a country where multiple serotypes circulate, the first two infections may cause illness, but the third and fourth tend to be asymptomatic. In children (and less commonly adults), the second infection may cause haemorrhage (DHF) and shock (dengue shock syndrome (DSS))
- it has an IP of 3–8 days.

Pathology and pathogenesis

Dengue is caused by an RNA flavivirus. DHF/DSS appears to occur as a result of non-neutralising antibody from previous exposure to a different serotype, enhancing uptake of immunocomplexes into monocytes. It is seen in infants who acquire dengue for the first time but who have circulating maternal antibody, and children during a second dengue infection. DSS is associated with cytokine release (especially TNF). DHF is associated with plasma leakage and haemorrhage.

Clinical features

Four syndromes may result from dengue virus infection: simple fever, dengue fever syndrome, DHF and DSS. The presentation is largely age dependent.

- Infants and young children infected for the first time tend to develop a simple fever
- Older children and adults infected for the first time develop dengue fever syndrome characterised by:
  - sudden onset of severe influenza-like symptoms with arthralgia (break-bone fever)
  - a maculo-papular rash on the 3rd day (in some) with petechiae developing later
  - severe retro-orbital pain on eye movement or pressure
  - occasional haemorrhage
  - prolonged convalescence
- Children previously infected with primary dengue and infected with

another serotype may develop DHF/DSS. In addition to the above, patients develop:

- serous effusions, especially pleural and peritoneal
- a haemorrhagic diathesis
- hepatomegaly and generalised lymphadenopathy
- hypotension and shock.

Diagnosis

Nonspecific features supporting a diagnosis of dengue fever are:

- leucopenia with relative lymphocytosis
- negative malarial films
- thrombocytopenia and clotting abnormalities
- hypoalbuminaemia.

Confirmation of dengue infection is by:

- virus isolation from the blood
- a four-fold rise or fall in IgG antibodies
- presence of IgM antibody.

Treatment

- For dengue fever: bed rest, sponging for fever, paracetamol (avoid aspirin because of the risk of haemorrhage)
- For DHF/DSS: intensive support is necessary.

Prevention

- Avoid mosquito bites by wearing appropriate clothing, using insect repellants, etc.
- Vector control: eliminating domestic breeding places and use of larvicides and insecticides
- There is no vaccine.

Prognosis

Uncomplicated dengue fever is a benign illness. The case fatality of DHF and DSS is 2 per cent.

Haemorrhagic fever with renal syndrome

Epidemiology

Hantaviruses are rodent viruses that cause a range of illnesses in humans.

- A family of hantaviruses exist, the members of which are associated with different clinical syndromes. Hantaan virus is the agent of haemorrhagic fever and renal syndrome (HFRS) which is distributed in

Russia, the Far East and Asia (reservoir: the field mouse); Puumula virus is the cause of nephropathia epidemica (a milder variant), distributed in Europe, especially Scandinavia (reservoir: the vole); and a new strain has been identified as the cause of a pulmonary syndrome (pulmonary syndrome hantavirus (PSHV)) identified in the USA (reservoir: the deer mouse)

- The incidence of HFRS is about 4/100,000 population in Russia
- Rodents excrete the virus asymptomatically. Human infection occurs via contamination of abrasions or mucous membranes by excretions, by inhalation of aerosols of dried saliva or excreta, or by rodent bites
- Most hantavirus infections are subclinical. Seroprevalence surveys show 10 per cent of persons in endemic areas have antibodies
- The IP is 2–3 weeks.

Pathology and pathogenesis

Hantaviruses belong to the Bunyavirus family, single-stranded RNA viruses. The acute renal disease is pathologically typical of an acute interstitial nephritis.

Clinical features

The clinical presentations range from a mild illness with minimal renal dysfunction, to one with severe renal failure, shock, pulmonary oedema and CNS involvement.

Classically, HFRS is divided into five clinical stages: febrile, hypotensive (day 5), oliguric (day 9), polyuric and convalescent (by day 14). Most patients present with:

- the abrupt onset of influenza-like symptoms
- blurred vision, conjunctivitis, flushed face, periorbital oedema and an erythematous rash
- palatal and axillary petechiae.

Complications

- Hypotension, shock
- Haemorrhage
- Renal failure
- Non-cardiogenic pulmonary oedema and ARDS (especially in PSHV)
- Encephalopathy.

Diagnosis

Suspicion should be aroused if there is:

- a history of exposure to rodents

- a combination of fever, thrombocytopenia, and acute renal failure or non-cardiogenic pulmonary oedema
- leucocytosis (>20.0 × 10^9/litre).

Confirmation is by:

- detection of IgM or a four-fold rise in IgG antibodies to a panel of hantaviruses
- detection of specific nucleic acid by amplification (PCR).

Treatment

- Ribavirin may be effective
- Intensive care and dialysis may be necessary.

Prevention

- Appropriate laboratory precautions should be observed when processing patient samples
- Reducing the risk of exposure to rodents
- No vaccine is available.

Prognosis

The fatality rate is approximately 5 per cent for HFRS, 1 per cent for nephropathia epidemica and >10 per cent for PSHV.

BRUCELLOSIS

Epidemiology

Brucellosis is a systemic bacterial infection caused by four different animal species: *B. abortus* (cattle), *B. melitensis* (goats), *B. suis* (pigs) and *B. canis* (dogs). Human brucellosis:

- accounts for 500,000 cases worldwide annually, with 20–30 cases in the UK every year
- is most commonly caused by *B. melitensis* (mainly Middle East, Mediterranean, Latin America, and Asia), followed by *B. abortus* (global)
- in the West is mainly imported in patients with a history of ingestion of unpasteurised dairy produce. Where *Brucella* is controlled in cattle, it is rarely acquired occupationally (e.g. farm workers, laboratory staff)
- is transmitted by ingestion of contaminated meat, milk or cheese, inhalation of infected material when handling animals (at parturition or slaughter), or inoculation through broken skin/mucous membrane contamination. Very rarely, infection may be acquired sexually, vertically or via blood transfusion

- is most common in young adults, and most severe when caused by *B. melitensis*
- may be subclinical, as identified by serological surveys
- has an IP from 1 to 3 weeks to several months.

Pathology and pathogenesis

Following ingestion or inhalation, the organisms reach the bloodstream via the lymphatics, causing bacteraemia. They are then taken up by macrophages, in which they multiply, and localise in the reticuloendothelial system. Once localised, they elicit a granulomatous reaction similar to that of tuberculosis, which may progress with tissue necrosis to abscess formation.

Clinical features

Onset may range from 1 day to 4 weeks. Typically, a patient with an acute or subacute illness may present with:

- influenza-like symptoms with drenching sweats, high hectic fever, rigors, myalgia and malaise. Recurring fever may develop (undulant fever)
- arthritis or arthralgia, usually monoarticular and large joint, and/or low back pain with sciatica
- headache, irritability, insomnia and confusion
- hepatosplenomegaly and lymphadenopathy.

Complications

- Arthritis, sacroileitis, vertebral osteomyelitis
- A toxic course with haemorrhagic features
- Meningitis, encephalitis and peripheral neuritis
- Endocarditis, myocarditis, pericarditis
- Granulomatous hepatitis
- Epididymo-orchitis
- Bronchopneumonia, pleurisy.

Bone and joint involvement occur in 10–30 per cent of cases. A chronic form of brucellosis has been described, with low-grade fever, fatigue, depression, insomnia and vague rheumatic symptoms. This is usually chronic fatigue syndrome in a person who has at some time acquired asymptomatic brucellosis and has positive antibodies.

Diagnosis

Nonspecific features supporting a diagnosis of brucellosis include a:

- history of consumption of raw milk or unpasteurised cheeses
- leucopenia with relative lymphocytosis
- raised ESR

- mildly deranged liver function tests (LFTs).

Confirmation of brucellosis is through:

- culture of the organism from blood, bone-marrow, urine or biopsy specimens. The cultures need prolonged incubation (up to 6 weeks). A positive blood culture is found in 30–50 per cent of *B. melitensis* cases, but less frequently with *B. abortus*
- serological tests. The standard tests are the serum agglutination test (SAT), complement fixation test, anti-human globulin test, and either ELISA or radioimmunoassay (IgM and IgG). The presence of IgM antibody, a four-fold rise or fall between acute and convalescent sera, or a single high IgG titre, indicate active infection.

Treatment

Combination therapy for 6 weeks is necessary with a tetracycline and an aminoglycoside being the preferred drugs.

- Doxycycline is the preferred tetracycline in combination with either: gentamicin or streptomycin (especially for in-patients), co-trimoxazole, ciprofloxacin or rifampicin
- A combination of co-trimoxazole and rifampicin is recommended during pregnancy.

Defervescence occurs within 4–5 days. A transient Herxheimer-like reaction may occur.

Prevention

- Skin testing cattle with *Brucella* antigen, and slaughtering those reacting, has eradicated *B. abortus* from cattle in the UK; however, this is impractical in many areas of the world
- Pasteurisation of milk meant for human consumption
- Education of those at risk, and the wearing of protective clothing, etc.
- Vaccination of young animals.

Prognosis

Most human infections are mild or subclinical, and are self-limiting over 2–3 weeks.

RICKETTSIAL DISEASES

Epidemiology

Rickettsia have mammalian reservoirs (humans included), are transmitted by arthropods (ticks, mites, lice or fleas) and are distributed in geographically distinct areas. The major *Rickettsia* are *R. prowazeki* (louse-borne (epidemic) typhus), *R. typhi* (murine (endemic) typhus), *R. tsutsugamushi* (scrub typhus), *R. rickettsii* (Rocky Mountain spotted fever (RMSF)), *R. conori* (tick typhus) and *R. akari* (rickettsial pox).

Louse-borne typhus:

- may cause epidemics in poor socioeconomic conditions
- is a disease of communities that do not change their clothes (e.g. cold, poor, war, famine; now a disease of tropical highlands); it is mostly reported from Africa
- is transmitted by the body louse (*Pediculus humanus*) when louse faeces infected with *R. prowazeki* contaminate the bite wound or mucous membranes
- has a reservoir of infection in humans (epidemics) and flying squirrels (sporadic cases)
- may recrudesce as Brill–Zinsser disease.

Murine typhus:

- is an urban zoonosis with a reservoir in rats
- is transmitted by the tropical rat flea (*Xenopsylla cheopis*) when flea faeces infected with *R. typhi* contaminate the bite wound or mucous membrane
- is endemic in southern USA, the Mediterranean, the Middle East, the Indian subcontinent, and South-East Asia.

Scrub typhu:

- is transmitted by the bite of the larval stage of trombiculid mites
- is widely distributed in the Asiatic–Pacific region
- is a rural disease
- is transferred transovarially in the mites
- commonly infects small mammals.

Spotted fevers (RMSF, tick typhus):

- can be caused by a range of *Rickettsia*, depending on the location: *R. rickettsii* (North and South America) to *R. conori* (Africa, Mediterranean)
- are transmitted by ticks whose bites often go unnoticed.

The IP for the rickettsial infections varies between 1 and 2 weeks, but occasionally may be as short as 2 days or as long as 18 days.

Pathology and pathogenesis

The prime pathological abnormality is a widespread vasculitis caused by rickettsial invasion of endothelial cells. All organs may be involved, but particularly the skin, brain, kidneys and heart. This leads to impaired vascular integrity and increased permeability giving rise to oedema, hypovolaemia, hypotension and hypoalbuminaemia. Perivascular inflammation, endothelial cell proliferation, arterial wall necrosis and thrombotic occlusion of arterioles occur.

Clinical features

The illness may pass unnoticed or be ascribed to a viral infection. The typical features of rickettsial disease are:

- abrupt onset of malaise, headache, fever, rigors and vomiting
- an eschcar at the site of the biting vector (scrub and tick typhus (50 per cent) and rickettsial pox (95 per cent)). It is a painless, punched-out ulcer covered with a black scab. There may be associated tender lymphadenopathy
- a maculo-papular or petechial rash. Characteristic features of this rash are as follows:
  - it is most prominent on the extremities in RMSF and tick typhus where it affects the palms and soles, and on the trunk in scrub, endemic and epidemic typhus
  - it develops from being maculo-papular into a petechial/purpuric rash (except in scrub typhus); in severe cases it is accompanied by ischaemia of the extremities
  - it is vesicular in rickettsial pox
  - it usually develops on day 3–5.

Complications

- Aseptic meningitis, encephalitis, optic neuritis, deafness
- Renal failure
- Non-cardiogenic pulmonary oedema, myocarditis, pneumonitis
- Recrudescence (*R. prowazeki*).

Diagnosis

RMSF and tick typhus have to be distinguished from measles, meningococcal septicaemia, vasculitis and idiopathic thrombocytopenic purpura, and early typhus from typhoid, malaria and relapsing fever. Nonspecific features supporting a diagnosis of rickettsial infection are:

- potential exposure to a vector
- a characteristic rash
- normal white cell count (WCCt) and ESR.

Confirmation of the diagnosis is by:

- direct immunofluorescence of a skin biopsy (RMSF)
- serology on acute and convalescent sera showing a four-fold rise in IgG antibodies or detection of IgM antibody
- isolation of *Rickettsia* from blood. This is hazardous, so is rarely attempted
- nucleic acid amplification (PCR).

Treatment

Rickettsia are insensitive to cell-wall antibiotics such as penicillins and cephalosporins. Active antibiotics include tetracycline (or doxycycline), chloramphenicol or ciprofloxacin.

The patient usually improves rapidly with appropriate therapy, with defervescence within 2 days. Intensive care is necessary for patients with severe vasculitis.

Prevention

- Vector (e.g. delousing with DDT) and reservoir (e.g. rat-curbing measures) control
- Personal measures to reduce insect bites, such as appropriate clothing and insect repellants
- Chemoprophylaxis with weekly doxycycline works in protecting against scrub typhus infection
- There are no vaccines available.

Prognosis

Epidemic louse-borne typhus and RMSF are the most serious. If antibiotic therapy is started before complications have occurred, fatalities are rare.

LEPTOSPIROSIS

Epidemiology

Leptospirosis is a zoonosis caused by a spirochaete with worldwide distribution. The genus *Leptospira* has been divided into two species: pathogenic (*L. interrogans*) or free-living (*L. biflexa*).

- 25–50 cases occur in the UK annually
- 202 pathogenic serovars of *L. interrogans* exist worldwide; 16 have been isolated in the UK
- Severity of illness is related to serovar, but not absolutely (e.g. Weil's disease is usually caused by *L. icterohaemorrhagiae*)
- Animals form the natural reservoir and rarely suffer disease
- The predominant UK strains are *L. icterohaemorrhagiae* (rat) and *L. hardjo* (cow)
- Pathogenic leptospires can live for 4 weeks in fresh water, 6 months in urine-saturated soil and 24 hours in seawater
- Transmission is when damaged skin or mucous membranes come into contact with infected animal urine
- Exposure usually occurs as a result of occupation (e.g. farm workers) or recreational pursuits (e.g. water sports)

- Most cases occur in working men and are reported during summer and autumn
- Serological surveys of at-risk populations show it to be an infrequent infection
- The IP is 10 days (range 5–19 days).

Pathology and pathogenesis

Leptospiraemia is followed by localisation in many areas, including the CSF. The disease is a vasculitis with capillary injury and haemorrhage resulting from immune-complex deposition. Jaundice is a result of hepatocellular dysfunction; biopsy shows no evidence of hepatocyte necrosis. Renal failure is primarily a result of tubular damage with acute tubular necrosis. During the leptospiraemic phase, CSF cultures are positive for leptospires but without any inflammatory reaction; this follows in the second phase when the cultures become negative.

Clinical features

The illness follows a biphasic pattern, the first phase representing the leptospiraemia and lasting 4–5 days. It is characterised by:

- abrupt onset of high fever, rigors and headache
- intense myalgia and muscle tenderness
- occasional confusion, abdominal pain, nausea and vomiting, cough, chest pains, maculo-papular rash and haemoptysis.

In many patients, the disease may not progress to a second phase and the patient recovers, having had an influenza-like illness. Features of the second phase include:

- recurrence of fever after 2–3 days
- suffused conjunctivae
- aseptic meningitis (Canicola fever) with a lymphocytic CSF, normal/raised protein and normal glucose
- deep jaundice, renal compromise and proteinuria (Weil's disease)
- pretibial raised erythematous rash (Fort Bragg fever).

Complications

- Renal failure
- Myocarditis
- ARDS and DIC with fulminant progression
- Chronic uveitis
- Relapse.

Most deaths occur around 14 days. Fulminant disease occurs rarely and is a multisystem disease requiring intensive care support. Myocarditis usually presents early with arrythmias. Anuria is common in Weil's disease, but is usually short-lived.

Diagnosis

The initial diagnosis is clinical, with the triad of high fever, renal failure and deep jaundice having few other causes. Nonspecific features supporting a diagnosis of leptospirosis include:

- the abrupt onset, severe myalgia and suffused conjunctivae
- a leucocytosis with a neutrophilia on differential WCCt
- a lymphocytic CSF
- a raised creatine phosphokinase (5 × normal in 50 per cent of cases)
- relatively normal LFTs. Typically there is a very high bilirubin level, with the transminases and alkaline phosphatase levels at between 1 × and 5 × normal (compared to 100 × normal in viral hepatitis)
- raised urea and creatinine, together with proteinuria.

Confirmation of leptospirosis can be demonstrated by:

- detection of specific antibody. This becomes positive by 7 days, the usual time of clinical presentation
- culture (takes weeks and is insensitive)
- identification of leptospires by dark ground illumination (technically difficult and false positive rates are high).

Treatment

All recognised cases should be treated. Mild infections recover without specific treatment and may only be recognised retrospectively. For severe or complicated disease, IV benzylpenicillin is the treatment of choice. It is effective up to the 2nd week of the illness. Tetracycline or doxycycline are alternatives. Dialysis is usually needed in patients developing renal failure.

Prevention

- Education of those at risk in reducing likelihood of infection
- Immunisation of domestic livestock and pets
- Prophylaxis with doxycycline if exposure is likely
- Control of the reservoir (e.g. rats)
- No human vaccine is available.

Prognosis

Death is very rare and is usually due to massive haemorrhage, acute renal failure or, occasionally, cardiac failure.

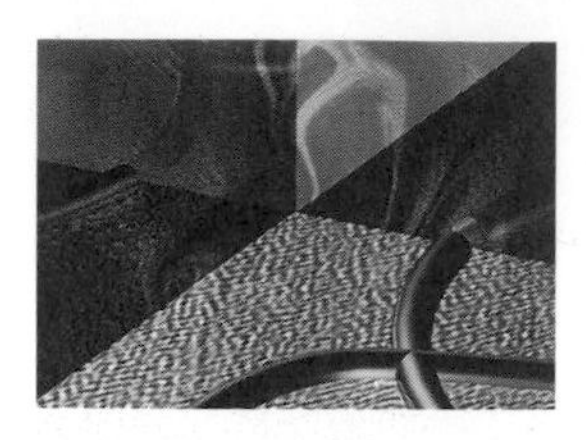

Helminthiasis

Helminths are parasitic worms. Of the many thousand varieties infecting all forms of animal and plant life, only a few are pathogenic.

THREADWORM (*ENTEROBIAS VERMICULARIS*)

Of worldwide prevalence, this is the most common worm infection in Britain, usually affecting children. Often several members of a household or institution are infected.

Cycle of infection

Adult worms are present in the colon and rectum, from where the gravid female emerges at night to deposit eggs in the perianal skin. These eggs are then carried by the hands, clothing or dust to be ingested by the same person or by new hosts. Swallowed eggs change to larvae in the intestine, maturing into adults which then breed, completing the cycle in about 1 month.

Clinical features and diagnosis

Perianal itching during egg deposition is the only symptom; white, slightly motile, gravid female thread-like worms may be visible in the faeces. 'Persisting' or relapsing infections are always due to reinfection, as adult worms can live only for 6 weeks. Confirmation is by microscopic demonstration of trapped eggs on a sticky cellophane tape applied to the perianal skin in early morning and then placed on a glass slide.

Treatment

Mebendazole, pyrantel and piperazine are all highly effective. Whole family or close group contacts should be treated at one time. Thorough cleaning of toilet and floors, laundering of clothing and bedding, continuing for several days after treatment, will remove eggs from the environment. Daily morning bathing, and hand washing after defaecation, discouragement of scratching the bare anal area and of nail biting are important.

HUMAN ROUNDWORM (*ASCARIS LUMBRICOIDES*)

This infection is very common in developing countries. Most cases occur in children.

Cycle of infection

Eggs are ingested via the hands, contaminated directly or indirectly by the faeces of another infected person. In the upper intestine, the eggs develop into larvae which penetrate the gut wall and proceed, via the liver, lungs, bronchus and trachea to the pharynx and are swallowed into the intestine, where they mature into large, pale pink, round worms 20–30 cm in length. Later they breed and produce a large number of eggs which are passed out with the faeces. Adults usually live for 12 months.

Clinical features and diagnosis

Many infections are asymptomatic. Less commonly, abdominal pain, urticaria, intestinal or biliary obstruction may occur. Visceral larva migration may cause tender hepatomegaly or allergic pneumonitis and an intense eosinophilia.

Passage of adult worms or the finding of ova in faecal smears or concentrate are diagnostic. Barium meal may reveal the worm (Fig. 18.1).

Treatment

Mebendazole, pyrantel and piperazine are all effective. Levamizole is very effective but is not currently available in the UK.

TOXOCARIASIS

This is caused by an inflammatory tissue reaction in humans provoked by larvae of animal roundworms *Toxocara canis* and *T. cati*, which normally infect dogs and cats. The adult worms are often found in puppies and pregnant bitches, and deposited eggs widely contaminate the soil. These eggs may be ingested by children and, in rare instances, the swallowed eggs may develop into larvae in the intestine, penetrate its wall and invade the liver, lungs, nervous system and eyes, causing eosinophilic

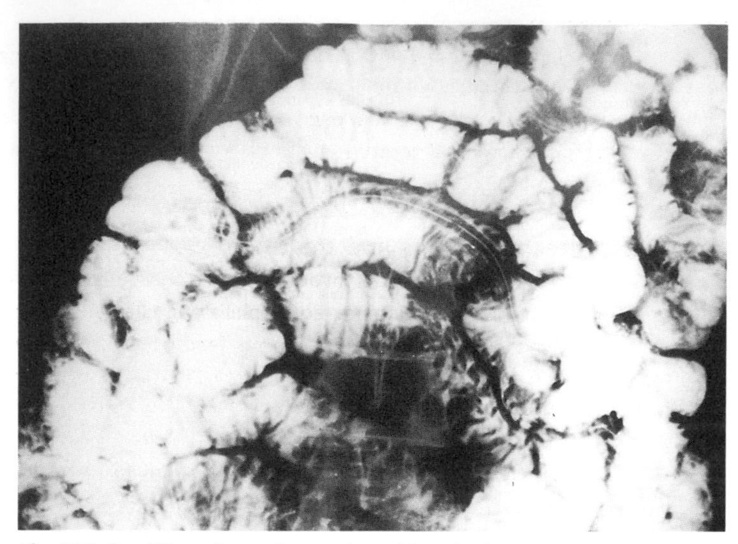

Fig. 18.1 Small bowel roundworm (*Ascaris*) on barium meal examination.

granulomatous reactions (visceral and ocular larva migrans). Larvae cannot complete the cycle of infection as humans are not the organism's natural host.

Clinical features, diagnosis and treatment

• There may be hepatomegaly, skin rash, choroidoretinitis, chest X-ray infiltration and visual impairment (this may occur years later)

• The diagnosis is suggested by the clinical features and eosinophilia, serology and histology (liver biopsy). Skin test helps confirmation. Treatment is largely supportive. Larvae can be killed by methylcarbamazepine but the usefulness of this is questionable.

Prevention

Keep dogs and cats out of children's sandpits and similar play areas.

HUMAN HOOKWORM

The main varieties of hookworm which infest humans are *Ancylostoma duodenale* (common hookworm) and *Necator americanus* (the 'American killer'). Infection is common in all tropical and subtropical countries where soil is widely contaminated with human faeces and where people walk barefoot.

Cycle of infection

Eggs deposited on soil in faeces develop into larvae, which penetrate the skin of bare feet and pass in succession, via lymphatics, blood stream, lungs and bronchi to the pharynx, where they are swallowed and fix to the wall of the small intestine, growing to the adult form. Egg production then takes place, completing the cycle (6–7 weeks).

Clinical features, diagnosis and treatment

At the time of larval entry through the skin there may be itching. Larval migration may cause intense eosinophilia and allergic pneumonitis (visceral larva migrans). The adult worms feed by sucking blood from the intestinal wall and, in a heavy infection, a progressively severe iron-deficiency anaemia develops. Light infections (less than 50 worms) usually do not cause anaemia.

Diagnosis is by the finding of ova in faecal smears or concentrates.

Mebendazole is an effective treatment (pyrantel is a suitable alternative).

Lightly-infected persons leaving an endemic country do not develop heavy infection later, as the worms cannot multiply within the intestine nor can the patient become reinfected from eggs in his own faeces.

CUTANEOUS LARVA MIGRANS

This is caused when a larva of an animal hookworm (usually *Ancylostoma braziliensis*, which normally infects dogs and cats) penetrates human skin. It then migrates slowly along the cutaneous lymphatics provoking a fierce tissue reaction which shows as a slowly advancing, serpiginous inflamed track on the skin (mostly of the feet). The larva cannot complete the life cycle as it is in the wrong host, and the condition is self-limiting within a few weeks. Thiabendazole or local freezing of the active end by carbon dioxide snow will expedite recovery.

TRICHINIASIS

This is a disease caused by the larvae of the intestinal nematode *Trichinella spiralis*, which infect and reside in the flesh, as encysted larvae, of a wide range of animals including domestic pigs, rats and wild animals. Ingestion of infected flesh leads to liberation of these larvae which mature into adults within the small intestinal mucosa. Breeding and larval production and invasion of flesh then follow. Undercooked pork and wild animal-meat are the main sources of human infection. There is a variable incidence worldwide dependent on animal rearing/feeding practices and wildlife meat eating.

Clinical features, diagnosis and treatment

- Initially there is abdominal pain and diarrhoea (during larval migration into intestinal mucosa), followed by systemic larval invasion; 1 week later muscle and joint pains and fever occur. After recovery, the visceral larvae are absorbed but the encysted muscle larvae may calcify after several years
- Intense eosinophilia is common during the acute stage. Serology is positive by the 3rd week. Muscle biopsy will show the encysted worms, often in considerable numbers
- Corticosteroids during the acute stage help, and thiabendazole is often used to kill the larvae.

Prevention

All fresh pork and wild animal meat should be adequately cooked. Prolonged freezing will destroy *Trichinella* cysts.

STRONGYLOIDES STERCORALIS

This small roundworm is found in all tropical and subtropical countries. It has two distinctive life cycles.

- In humans, the cycle is similar to that of hookworm, except that there is no male parasite and the female is parthenogenetic. Also, ova mature into larvae *within* the intestine rather than *outside*. This may lead to autoinfection
- Free-living life cycle in soil, involving both adult male and female parasites.

Clinical features, diagnosis and treatment

- Often asymptomatic, but in some cases larvae may penetrate the skin around the anus producing urticarial skin eruptions and migratory serpiginous rash extending rapidly over the thighs and trunk (larva currans). In immunocompromised hosts, intestinal autoinfection may lead to heavy worm-load, causing chronic diarrhoea and malabsorption; also disseminated disease with wasting and pulmonary involvement may develop, which may be fatal (Fig. 18.2)
- Eosinophilia is common during visceral migration and in chronic infection. Diagnosis is by demonstrating larvae in fresh stools, duodenal aspirate or sputum (rarely)
- Treatment is by thiabendazole or albendazole.

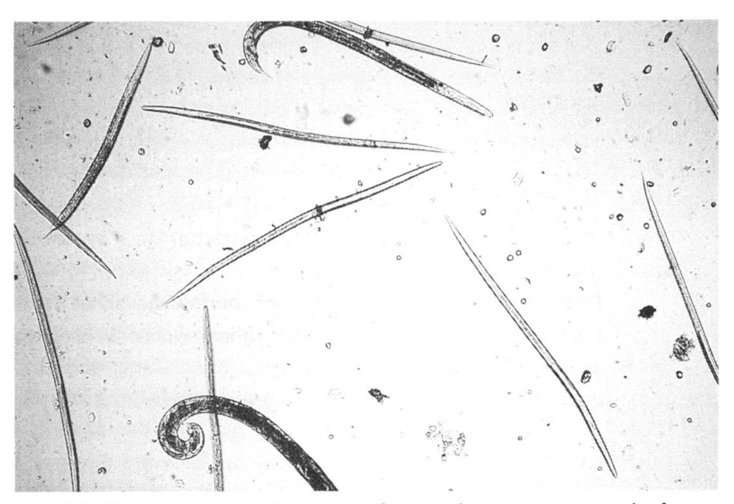

Fig. 18.2 *Strongyloides* larva in sputum from an immumocompromised person.

WHIPWORM (*TRICHURIS TRICHIURA*)

This is a small roundworm occurring in all warm countries with poor sanitary facilities. The cycle of infection is a simple transfer of ova by the bowel–soil–hand–mouth route. Light infections are symptomless and harmless. Rarely, a heavy infestation will cause abdominal pain, bloodstreaked diarrhoea and anaemia. The diagnosis is by the finding of ova in the faeces. The treatments of choice are thiabendazole or mebendazole.

TAPEWORMS (CESTODES)

Tapeworms are elongated flatworms which may reach 10 m in length. They are hermaphrodite, so a single adult worm may produce huge numbers of fertile eggs.

Life cycle

This involves two vertebrate hosts: an intermediate and a definitive.

The intermediate host ingests ova which develop into larvae and penetrate the host animal's muscles and other organs.

The definitive host acquires the infection by ingestion of the raw flesh of an intermediate host. In the gut the larva matures into an adult worm whose head attaches to the upper intestinal mucosa. The body of the worm consists largely of multiple, flat, egg-bearing segments which detach from time to time and shed in the faeces. The ova and segments in the faeces infect new intermediate hosts, completing the double cycle of infection.

Humans may be involved as definitive or intermediate hosts.

HUMAN DEFINITIVE HOST INFECTIONS

Epidemiology and clinical features

Fish tapeworm (*Diphyllobothrium latum*)

Humans acquire infection with the fish tapeworm by the ingestion of raw freshwater fish. The infection is common in Scandinavia and the Far East, but is not found in Britain. The infection is generally symptomless, but if the worm fixes at the upper end of the jejunum it may successfully compete for the host's dietary vitamin B_{12} and a megaloblastic anaemia may occur. The diagnosis is made by the finding of ova in the faeces.

Beef tapeworm (*Taenia saginata*)

Humans are the only definitive hosts of this worm, and they acquire the infection by eating the undercooked flesh of the cow, the common intermediate host. The infection is usually symptomless but may cause great anxiety to the patient, as the detached, whitish-yellow, opaque segments are motile and may emerge spontaneously from the anus. Infection in Britain is uncommon but is common in tropical countries. Infected carcasses may be detected in slaughterhouses by examination of the masseter muscles.

Pork tapeworm (*Taenia solium*)

Humans are the only definitive hosts, and they acquire the infection by eating undercooked 'measly' pork or raw pork sausages. The adult worm infection usually causes no symptoms other than vague abdominal complaints.

Diagnosis and treatment

• Infection with beef or pork tapeworms is diagnosed by the identification of the segments or ova in faeces
• Differentiation between the two can only be made by examining the morphology of the head (scolex) or egg-bearing segments (proglottids)
• All intestinal tapeworms are readily destroyed by niclosamide (Yomesan). A saline purge 1 hour after treatment should be given in *T. solium* to prevent regurgitation of eggs into the stomach, which may lead to cysticercosis. Occasionally, worms are not eradicated, ova or seg-

ments reappear in the faeces and retreatment is necessary. Identification of the head in the faeces after treatment confirms cure.

Prevention

Measures include the treatment of human cases, treatment of raw sewage, inspection of meat and pork in slaughterhouses, adequate cooking of fish, beef and pork, and deep-freezing which kills the larval worms.

HUMAN INTERMEDIATE HOST INFECTIONS

Cysticercosis (*Taenia solium*)

Humans become infected as an intermediate host either by the ingestion of ova from another human host or from their own intestines. The ova develop into larvae which penetrate the gut wall and invade the tissues, particularly the skeletal muscle and the brain, where they mature to cysticercus forms. During the invasive stage there may be slight fever, muscle aching and eosinophilia. After a latent period of several years the cysticercus forms die, producing inflammatory and fibrotic lesions which eventually calcify (rarely in the brain: Fig. 18.3). At this stage, recurrent grand mal convulsions or, occasionally, more diffuse neurological, ocular or mental disturbances may occur.

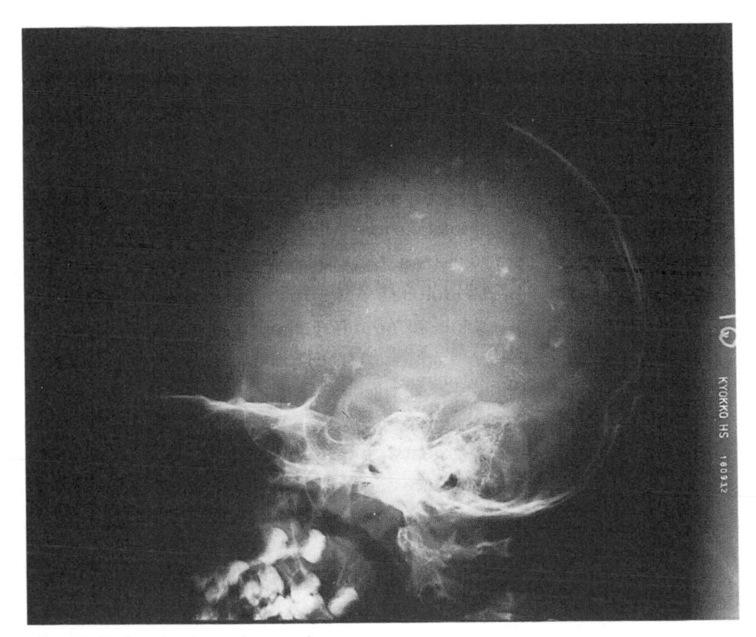

Fig. 18.3 Cerebral cysticercosis.

Diagnosis

The diagnosis is suspected in a patient who develops recurrent fits after living in the tropics for a period, and may be confirmed by the finding of numerous calcified cysts on muscle X-ray and by serological tests. Computerised tomography (CT) or magnetic resonance (MR) scan of the brain show cystic or solid nodular or calcified lesion.

Treatment

Anti-convulsant drugs are the mainstay of management, but praziquantel may be used if there is suspicion of live infection. Symptoms may exacerbate during therapy and corticosteroids should be used to prevent this.

Hydatid cysts

The dog is the definitive host of the small tapeworm *Taenia echinococcus* which produces large numbers of eggs in the faeces, contaminating grass and infecting sheep, which are the usual intermediate hosts. The cycle is completed by the ingestion of infected raw sheep offal by dogs.

Humans become accidental intermediate hosts by close contact with dogs. The disease is particulary prevalent in the Middle East, but occurs in all sheep-rearing countries and has been reported in Wales. In humans, the ingested ova develop to form embryo worms which penetrate the gut wall and invade the tissues. Most of the embryos are destroyed, but

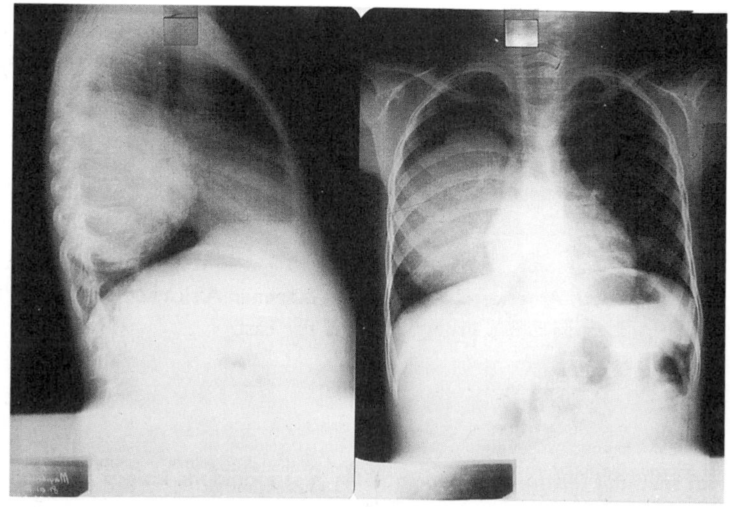

Fig. 18.4 Pulmonary hydatid cyst in a young shepherd.

one or more may survive and become encysted, usually in the liver, occasionally in the lung or rarely in other organs. These hydatid cysts gradually enlarge over a period of years, eventually reaching 5–20 cm in size. The fluid-filled cyst has a thick capsule and contains numerous infective protoscolices.

Clinical features

Clinically the cyst presents as a palpable, slowly growing liver tumour or as a partially calcified circular lesion on chest X-ray (Fig. 18.4). Large cysts in the liver may rupture spontaneously, causing sudden pain, fever, allergic rashes and eosinophilia. Slow development of numerous daughter cysts follows in the contaminated areas. Cysts may calcify gradually with the death of all living worm tissue.

Diagnosis

The diagnosis is usually by ultrasound or CT scans. Serological tests will provide confirmation.

Treatment

Albendazole in three cycles of 28 days each may succeed in killing infective cysts. Hepatic hydatic cysts often grow very slowly and remain asymptomatic for years, and may not require treatment unless when very large or causing symptoms due to compression. Operative removal of the cyst under a pre- and post-albendazole treatment regimen is the mainstay of therapy. Calcified cysts should be left alone.

SCHISTOSOMIASIS (BILHARZIASIS)

Schistosomiasis is caused by the blood fluke, *Schistosoma*. The three major species of human importance are *S. mansoni*, *S. haematobium* and *S. japonicum*.

Epidemiology

Schistosoma mansoni is endemic in Africa, the Middle East and parts of South America, *S. haematobium* is seen in Africa and the Middle East, and *S. japonicum* is prevalent in the Far East.

Life cycle and transmission

Humans are the principal hosts and freshwater snails of particular types are the intermediate hosts. Adult males and females live within the venules of the human bladder and intestine for years and produce eggs, which reach fresh water sources via faeces or urine.

Larvae are liberated in water and penetrate into snails, where they mature and later emerge as free-swimming larvae (cercariae). Cercariae penetrate human skin in contact with water and reach the liver via the blood stream. Here they mature into adults and migrate into mesenteric (*S. mansoni* and *S. japonicum*) or bladder (*S. haematobium*) veins, when egg production occurs (about 6–12 weeks after infection).

Pathogenesis

Eggs are released into the lumen of the bowel or bladder and are also transported by blood to the liver, lungs and other sites with granuloma formation at the site of egg deposition (intestinal mucosa, urinary tract, liver, lungs, etc.) and fibrosis. Clinical disease results from repeated egg deposition in large numbers and requires repeated infections in endemic countries to produce a large worm-load. Most infections are asymptomatic.

Clinical features

Acute stage (Katayama fever): this is related to an immune reaction to initial worm maturation and egg deposition. There may be fever, urticaria, headache, abdominal pain and diarrhoea lasting for a variable period

Chronic intestinal and hepatic schistosomiasis (*S. mansoni* and *S. japonicum*)*:* occasional blood-stained stools due to the formation of large bowel inflammatory polyps; possibly colorectal cancer; hepatomegaly and splenomegaly due to hepatic fibrosis and portal hypertension

Chronic urinary schistosomiasis (*S. haematobium*)*:* dysuria, frequency and haematuria due to obstructive uropathy; bladder inflammatory polyp and secondary infection; possibly bladder cancer.

Pulmonary fibrosis and hypertension, myelitis (can occur during acute stage) and cerebral symptoms are other features.

Diagnosis

Demonstration of eggs in stool, urine or biopsy (rectal, bladder, liver) is required for definitive diagnosis. Remember egg deposition may take up to 12 weeks from infection. Serology is helpful in diagnosing recent infection or acute disease in a traveller.

Treatment and prevention

Praziquantel is the drug of choice for all three types of schistosomiasis. Important prevention measures are: education of public; sanitary disposal of faeces and urine; measures to reduce the snail population; provision of safe water for drinking and bathing; avoidance of swimming in fresh water in endemic countries. Treatment of infected people helps to reduce morbidity and infectivity.

FILARIASIS

Three main clinical forms of human filariasis are recognised: lymphatic filariasis, onchocerciasis and loaisis.

Lymphatic filariasis

This is usually caused by *Wuchereria bancrofti* which affects millions of people living in equatorial Africa, the Indian subcontinent, South-East Asia, Central and South America.

Life cycle

Humans become infected through mosquito bites. Injected larvae migrate to the lymphatics where they mature to the adult form, and are able to produce microfilariae in 6–12 months from infection (these periodically migrate to the peripheral blood and are taken up by biting mosquitoes).

Clinical features and diagnosis

- Infected persons may remain totally asymptomatic despite microfilaraemia, or suffer from episodes of fever, lymphangitis, lymphadenitis, epididymitis or orchitis; microfilaraemia is common (particularly late at night). Recurrent lymphatic inflammation may lead to chronic lymphatic obstruction producing thickened, oedematous, swollen hyperkeratotic skin, involving the legs, genitalia and breasts (elephantiasis). Secondary skin sepsis is common. Detectable microfilaraemia is rare
- Tropical pulmonary eosinophilia syndrome: recurrent asthma and fever (often nocturnal), associated with marked eosinophilia and diffuse lung infiltrates (on X-ray). Microfilaraemia is absent
- The early manifestations are usually diagnosed by demonstrating microfilaria in blood film. Chronic filariasis is a clinical diagnosis but filaria antibodies are usually present in blood. Tropical pulmonary eosinophilia is diagnosed by demonstrating high titre serum filarial antibodies in association with characteristic clinical, haematological and radiological findings.

Treatment and prevention

Diethylcarbamazepine rapidly eliminates microfilaria and symptoms of early disease. Adult worms are killed less easily and repeat courses are necessary. Febrile and local tissue reactions are common during treatment due to disintegrating parasites, and corticosteroids are helpful. Obstructive features of chronic disease may be helped by surgical reconstruction.

Vector control, insect repellents and use of mosquito nets and screens are important preventive measures.

Onchocerciasis

This is caused by *Onchocerca volvulus*, and is endemic in equatorial Africa and South America. The vector is the female blackfly.

Life cycle

Humans become infected through inoculation of larvae by a biting blackfly. Within the skin the larvae mature into adults after a few months. Several males and females lie coiled together in round bundles in subcutaneous tissue. Production of microfilariae begins with migration to the skin and and eyes. The female blackfly ingests microfilariae while biting the skin.

Clinical features and diagnosis

- Dermatitis: most common (later, the skin becomes thickened and wrinkled); visible or palpable subcutaneous nodules, enlarged lymph glands and visual impairment due to conjunctivitis, keratitis, anterior uveitis and choroidoretinitis
- Diagnosis is by demonstrating microfilariae in skin biopsy or cornea (slit lamp examination), or adult worms in excised subcutaneous nodules.

Treatment and prevention

Ivermectin as a single dose repeated yearly (until adult worms die) is the drug of choice. Use of protective clothing and headgear, insect repellants, vector control measures and treatment of infection to reduce the reservoir of infection are the main principles of prevention, but these are difficult to implement.

Loaisis

Epidemiology and pathogenesis

Loaisis is caused by adult *Loa loa* worms, and is seen mainly in central African rain forests. Vectors are horsefly or deerfly of the genus

Chrysops. Larvae injected into humans during the bite of an infected fly mature to adult worms, which live in subcutaneous tissue. Microfilariae produced later appear in the blood and are taken up by the biting flies.

Clinical features and diagnosis

- Clinical manifestations are the result of worm migrations. Typical are transient subcutaneous swellings. Distal extremities and the periorbital skin are favourite sites. Worms may migrate across the scleral or conjunctival tissue. Fever and urticaria may occur
- Diagnosis is by demonstrating microfilariae in blood film.

Treatment and prevention

Diethylcarbamazepine in repeat courses may be necessary. There may be severe and systemic reactions and in-patient supervision is recommended.

Insect-repellant measures should be adopted.

Index